Gas Flow Modeling with Shock Waves Using the Gas-Hydraulic Analogy

— An Illustrated Album —

Joseph Yakubov

Bakhram Musabekov

Tashkent – Providence

2025

Silk Road
Memoirs

Gas Flow Modeling with Shock Waves

Using the Gas-Hydraulic Analogy: An Illustrated Album

 ISBN: 979-8-9992216-9-8

Published in the United States by Silk Road Memoirs

Printed by Amazon KDP, www.kdp.amazon.com

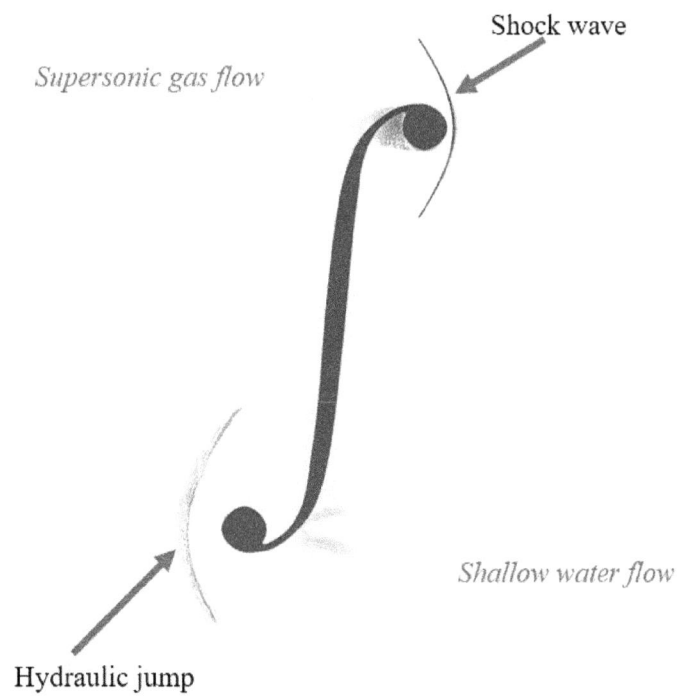

Supersonic gas flow

Shock wave

Hydraulic jump

Shallow water flow

The emblem of the gas-hydraulic analogy method, designed in 1977, symbolizes the conceptual and mathematical correspondence between compressible gas flow and shallow water flow.

Contents

Introduction

"We who work in fluid mechanics are fortunate, as our colleagues in a few other fields such as optics, that our subject is easily visualized."

This is the opening line of *An Album of Fluid Motion* compiled by Milton Van Dyke, a book that nearly every fluid scientist has leafed through in admiration of its striking flow images. Yet, among its many unforgettable photographs, only one—Image No. 229, titled *Hydraulic analogy for a wedge in supersonic flow*—depicts the gas-hydraulic analogy method. Unfortunately, the chosen image is rather unconvincing and inadvertently casts the method in a poor light.

Even the *Gallery of Fluid Motion*, an annual showcase launched in 1983 by the Division of Fluid Dynamics of the American Physical Society (APS) and inspired by Van Dyke's work, has consistently overlooked the gas-hydraulic analogy method.

This photo album aims to correct that omission. It presents rare and compelling visualizations that demonstrate the capabilities of the gas-hydraulic analogy—a method based on the mathematical similarity between the shallow water equations and the equations of motion for an ideal compressible gas.

The images illustrate flow phenomena that are difficult or impossible to model using conventional tools such as supersonic wind tunnels, shock tubes, or ballistic installations.

The advantages of the hydraulic analogy are significant:

- Thousand-fold time dilation of the observed processes, making them highly accessible and effective for education.

- Possibility of modeling complex non-stationary problems of aerodynamics, difficult or impossible to solve by other methods of aerodynamic experiment, and at the same time, while also allowing the identification and analysis of hysteresis effects.

- Substantial savings in both cost and time during the early stages of aerodynamic research and design.

A theoretically inclined academic might dismiss these as merely two-dimensional problems solvable by modern computation. But an aerospace engineer, engaged in the development of a new flight vehicle, will recognize the value in visually exploring the flow structures that may arise around their design.

All photographs in this album come from the personal archive of Joseph Yakubov, who led research in gas-hydraulic analogy at the Tashkent Polytechnical Institute named after Abu Rayhan Beruni from the 1960s until the collapse of the Soviet Union.

Experimental Apparatus

Photo 1.

The Gas-Hydraulic Analogy makes it possible to study various cases of gas motion, including the flow around bodies placed in a gas stream, using hydraulic models. For this purpose, analog devices known as shallow water channels or hydraulic channels are used.

The most effective method involves towing the test models through shallow water — that is, the medium (water) remains stationary while the body moves.

This is referred to as direct motion modeling.

Technical Specifications of the HC-1 Analogy Device (Tashkent):

1. Flume dimensions:

Length — 11 m; Width — 1.25 m

2. Adjustable water depth range:

From 5 mm to 20 mm

3. Achievable range of analog Mach numbers (Froude numbers):

From 0.6 to 2.5
(Corresponding to model towing speeds in still water from 0.1 to 1.0 m/s)

4. Range of model lengths (streamwise dimension):

From 150 mm to 400 mm

5. Drive power of the towing carriage:

0.3 kW

6. Model movement control:

Remote-controlled, with real-time monitoring via a video system for visual observation of the flow processes.

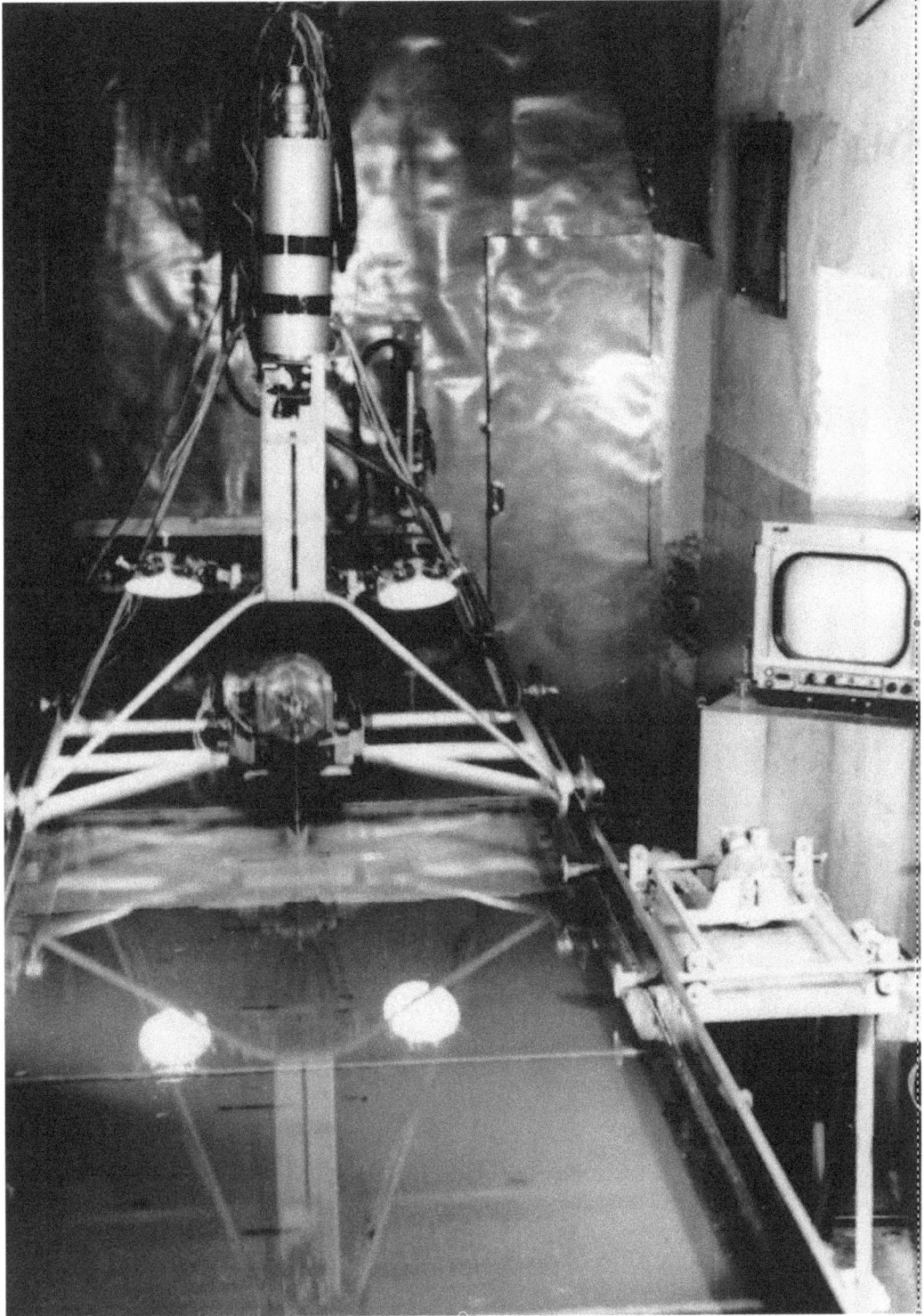

Photo 1
Analog Device HC-1
(Shallow water hydraulic channel)

Photo 2

Photo 3, 4

HC-2 is a complex system that includes the following components:

1. **Hydraulic Channel**
   (a stationary shallow-water flume consisting of two sections — one circular and one rectangular in plan), as well as:

   - a mobile platform moving at a preset speed along rails;
   - a rotational device;
   - a radial carriage with a wave generator.
   - a drive system for the moving platform;
   - imaging and lighting systems;
   - a coordinate device for model control;
   - and a central control console.

2. **Remote Control System**
   Provides both manual and automatic control of model movement.
3. **Measurement and Recording System**, including:

   - a video observation unit;
   - and a laser interferometric system for registering wave structures.

4. **Data Processing System**
   Used for analyzing the results of analog experiments.

Technical Specifications

1. **Geometric Parameters:**

   - Diameter of the circular section: 5 meters
   - Width of the rectangular section: 3 meters
   - Total channel length: 10.7 meters

2. **Water Depth Range:**

   - In shallow-water modeling mode: 6–20 mm
   - In hydrodynamic similarity mode: 40–100 mm

3. **Model Size (along flow direction):**
   From 150 mm to 500 mm
4. **Range of Simulated Analog Mach Numbers:**
   From 0.6 to 4.0
5. **Drive Power of the Moving Platform:**
   1.8 kW

Photo 3
Analog device HC-2
(Shallow water hydraulic channel)

Photo 4

Analog device HC-2

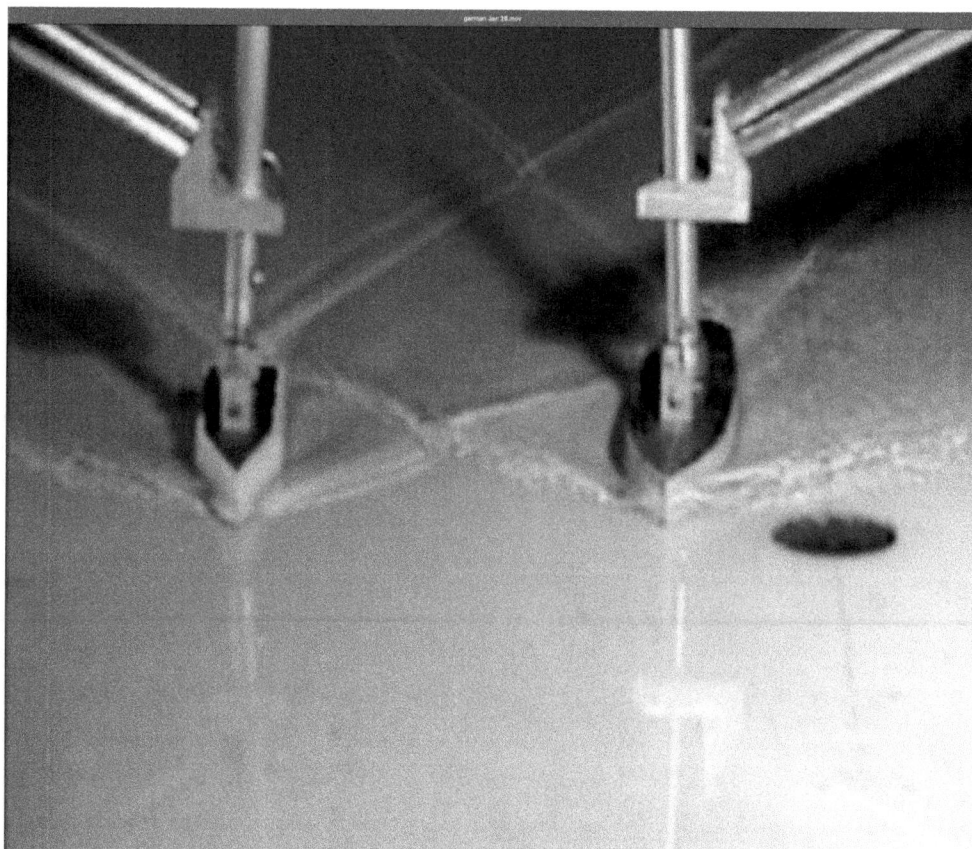

Photo 5

Translational Motion of Models Along the Hydraulic Channel HC-2

Hydraulic Analogy of Supersonic Flow over a Wedge

(based on Klein, 1965)

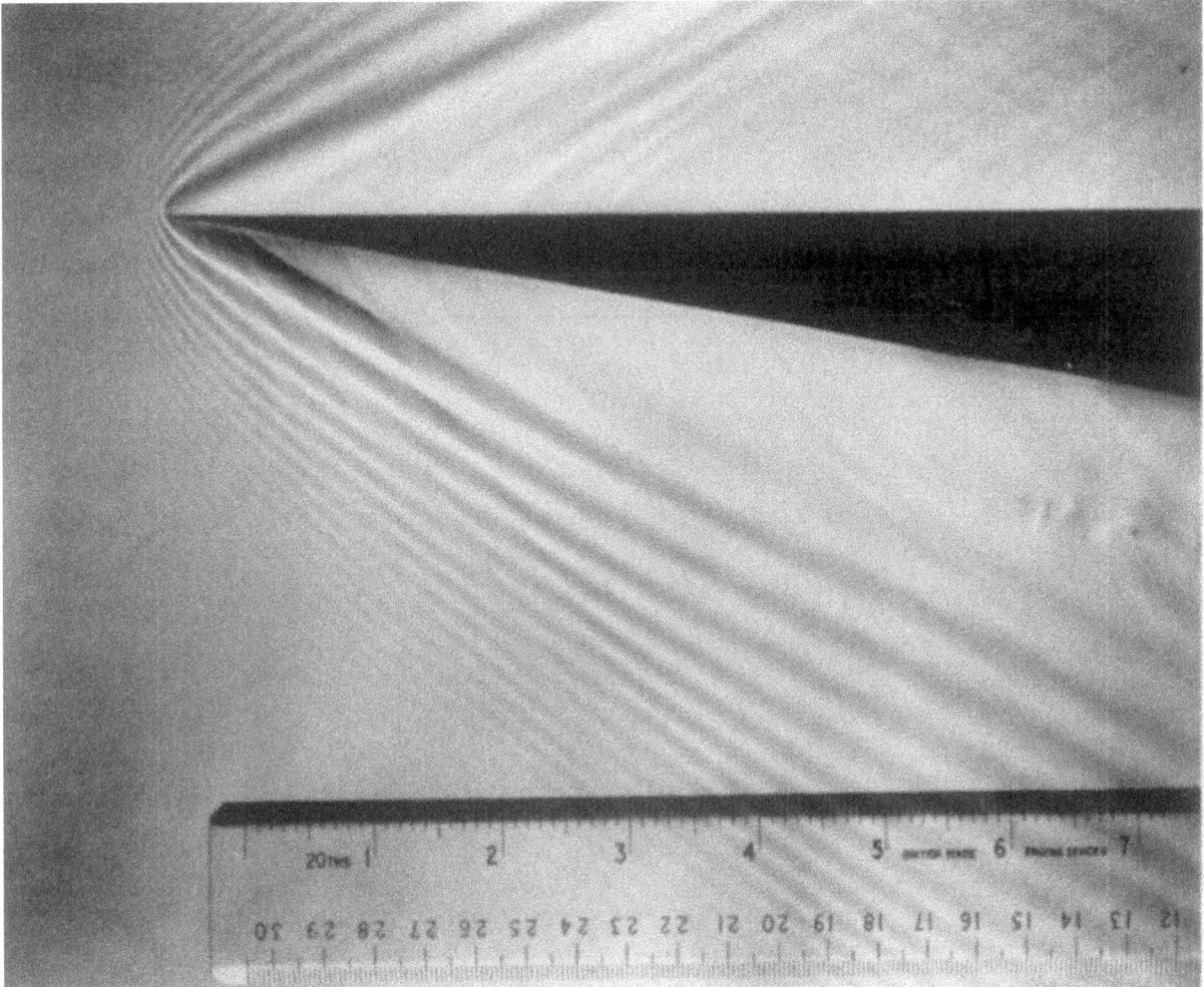

Photo 6

"Fig. 5.6 Hydraulic analogy to the supersonic flow over a wedge. (From Klein, 1965.)" [7]. In Milton Van Dyke's classic book, a similar image attributed to Klein—taken from Merkitz—is included, but it is poorly executed and, rather than illustrating the method's potential, inadvertently serves as a negative example.

Photo 7

Hydraulic analogy to the supersonic flow over a wedge. (From HC-1)

Comparison of Results: Gas-Hydraulic Analogy vs. Wind Tunnel Experiments

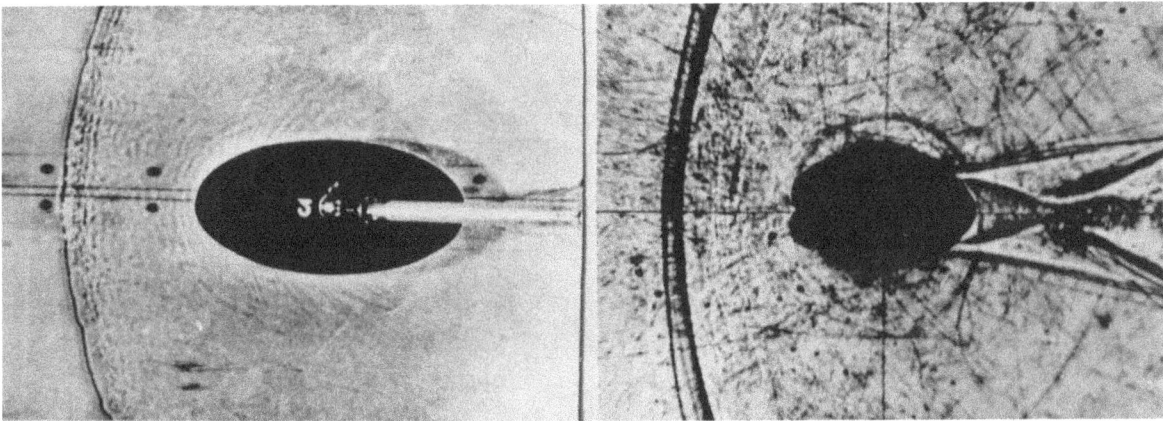

Photo 8.
Comparison of shadow patterns in supersonic
flow around an elliptical cylinder (Mach number $M_\infty = 1.5$).
Left: obtained using the hydraulic analogy device HC-2.
Right: photograph taken in a supersonic wind tunnel.

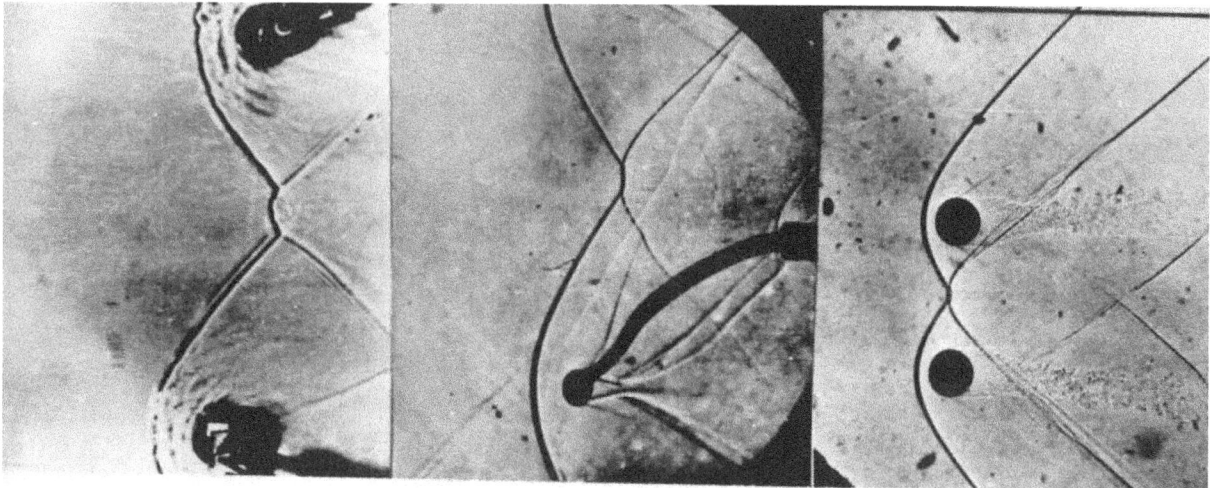

Photo 9.
Flow around two cylinders.
Left: Hydraulic analogy visualization using device HC-1
(Tashkent Polytechnical Institute named after Abu Rayhan Beruni)
Center: Supersonic wind tunnel visualization
(Institute of Mechanics, Lomonosov Moscow State University, Moscow).
Right: obtained using a ballistic installation
(A.F. Ioffe Institute, St. Petersburg).

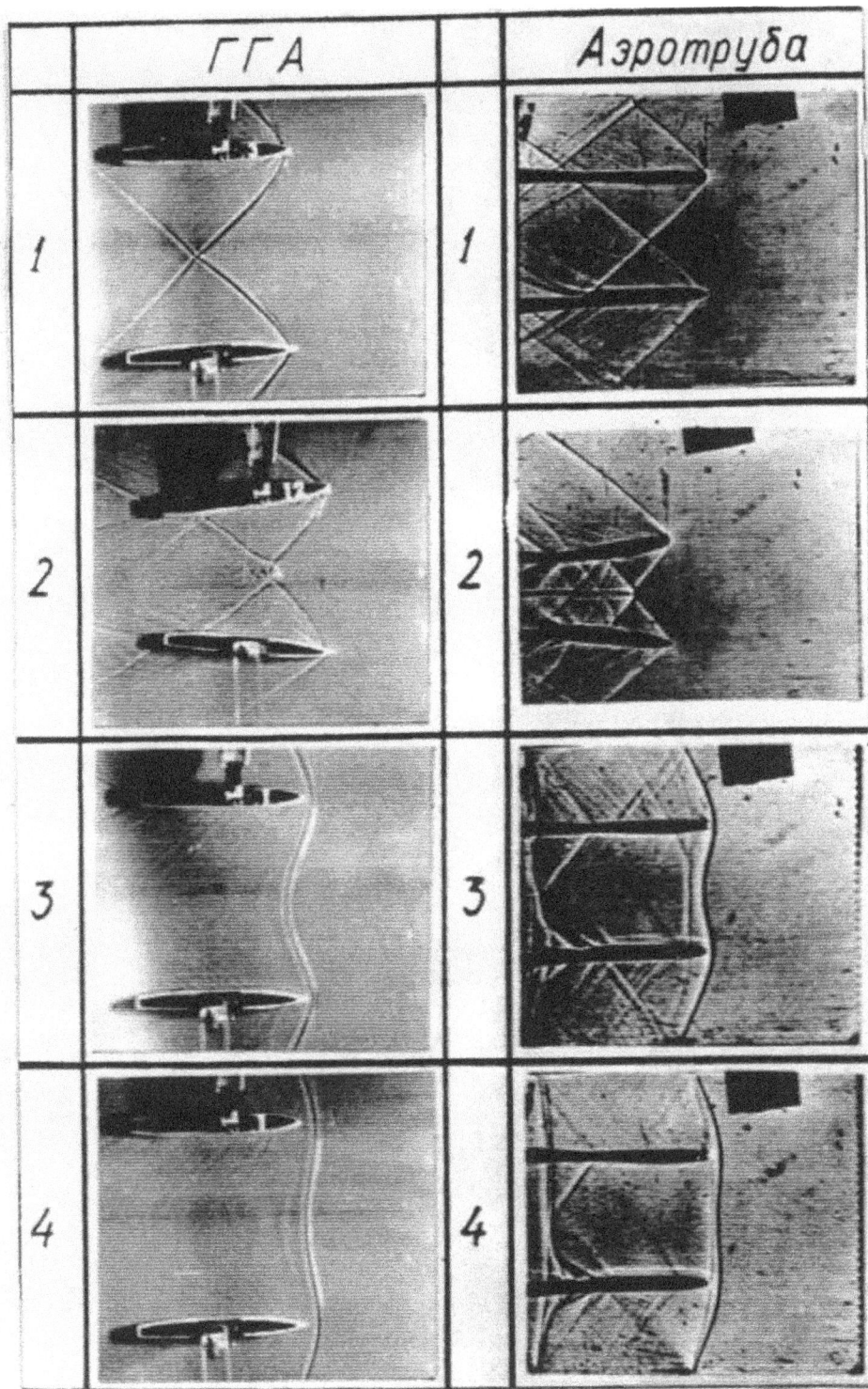

Photo 10
Left: photographs taken in the HC-1 hydraulic channel; right: corresponding images obtained in a supersonic wind tunnel (A. I. Shvets, Institute of Mechanics, Lomonosov Moscow State University).

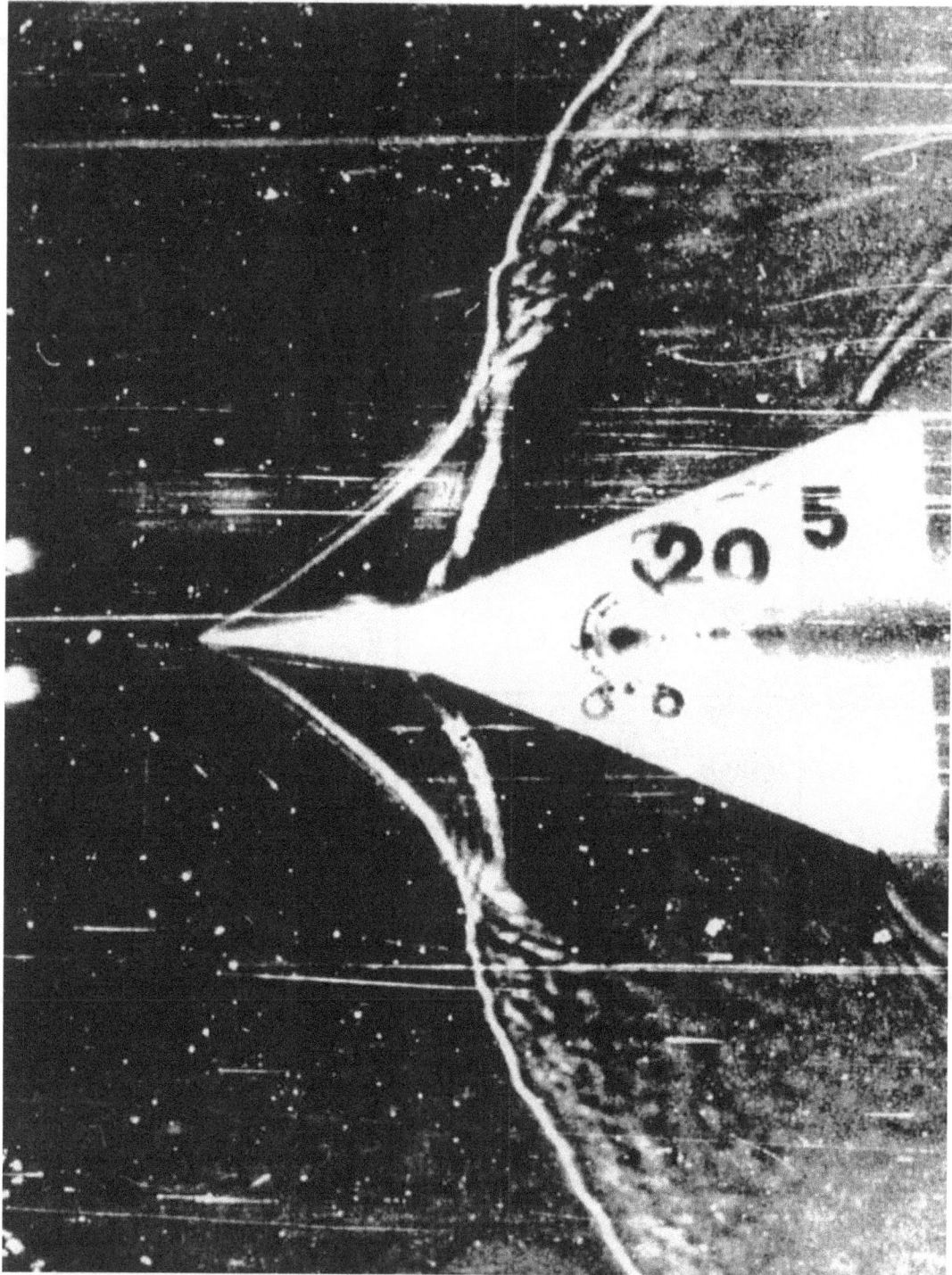

Photo 11.
Hydraulic analogy visualization of
supersonic flow ($M_\infty = 2.05$) around a double
wedge. Obtained using device HC-1 at the
Tashkent Polytechnical Institute named after
Abu Rayhan Beruni

Photo 12

Regular reflection of an incident shock

wave from a round convex surface

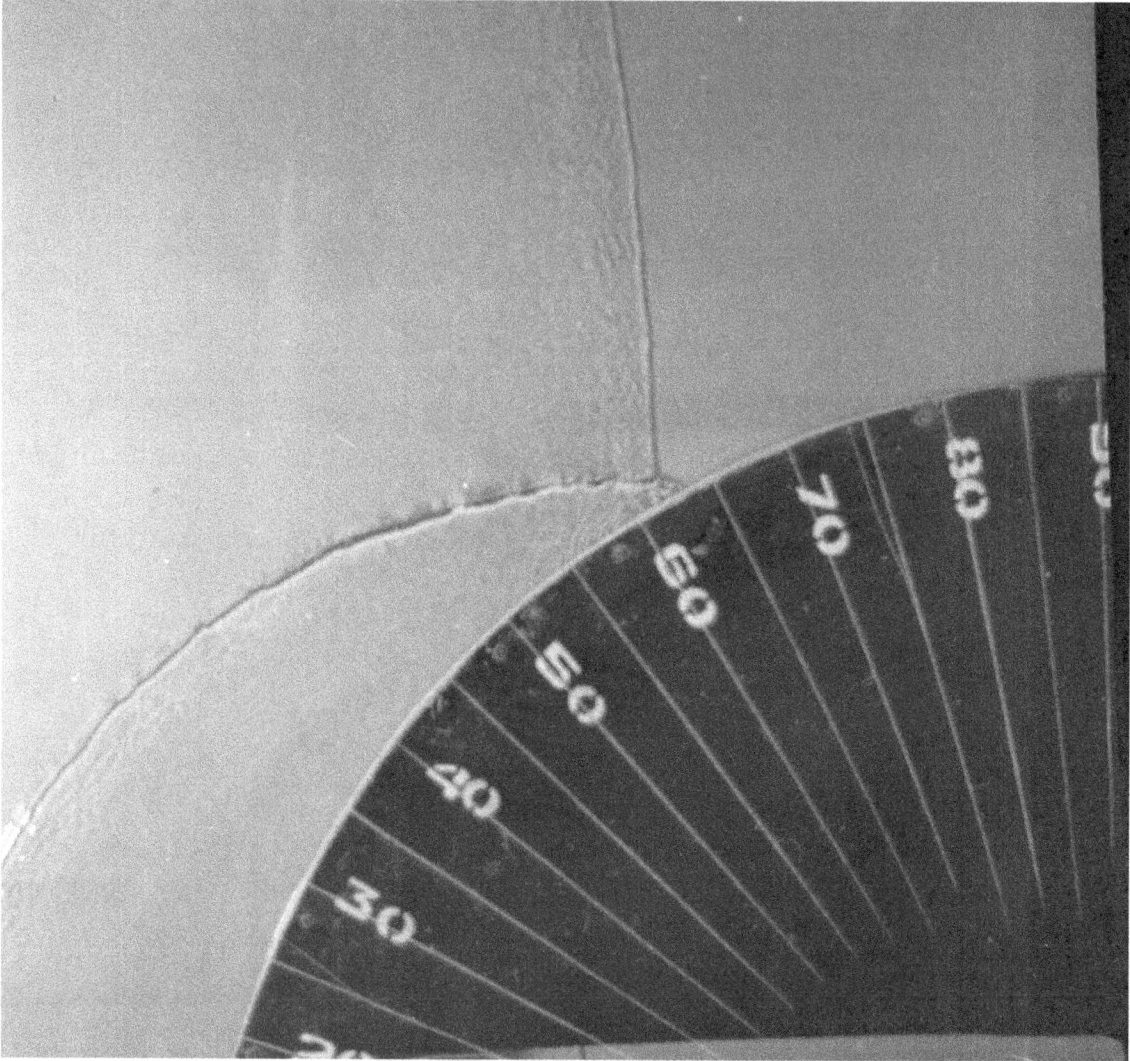

Photo 13

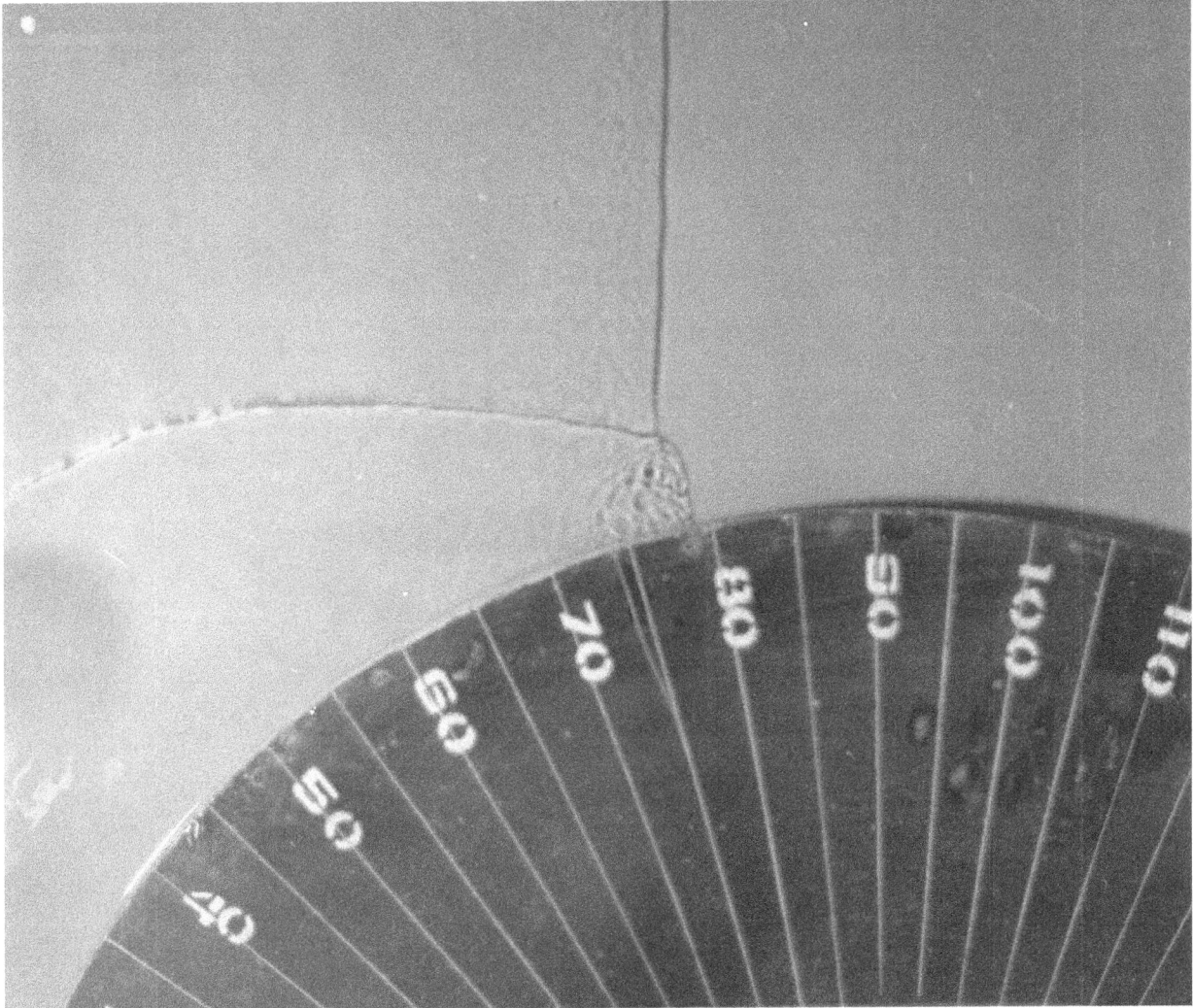

Photo 14

The incident and reflected
shock waves, Mach stem, and tangential
discontinuity are shown on the round
convex surface.

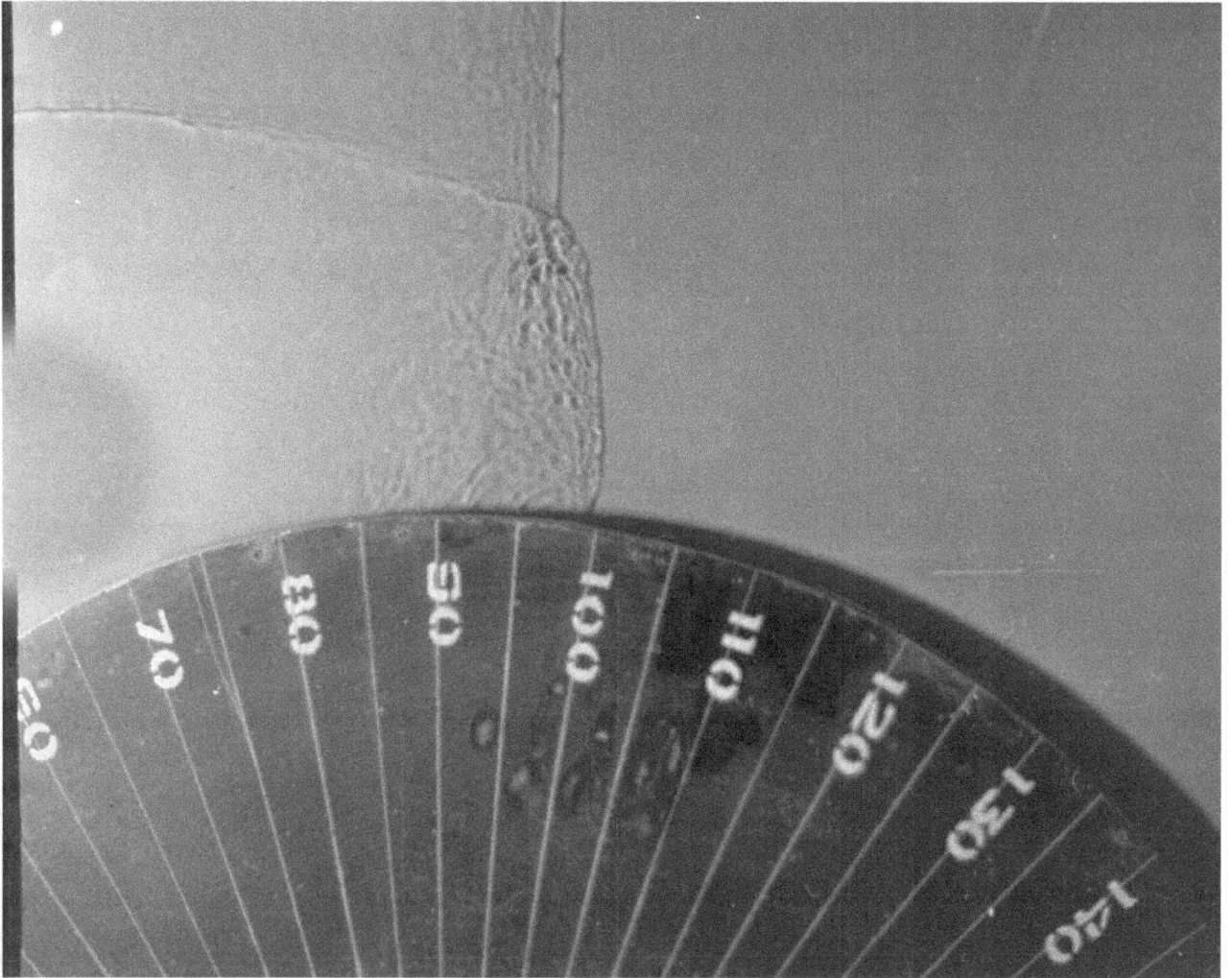

Photo 15

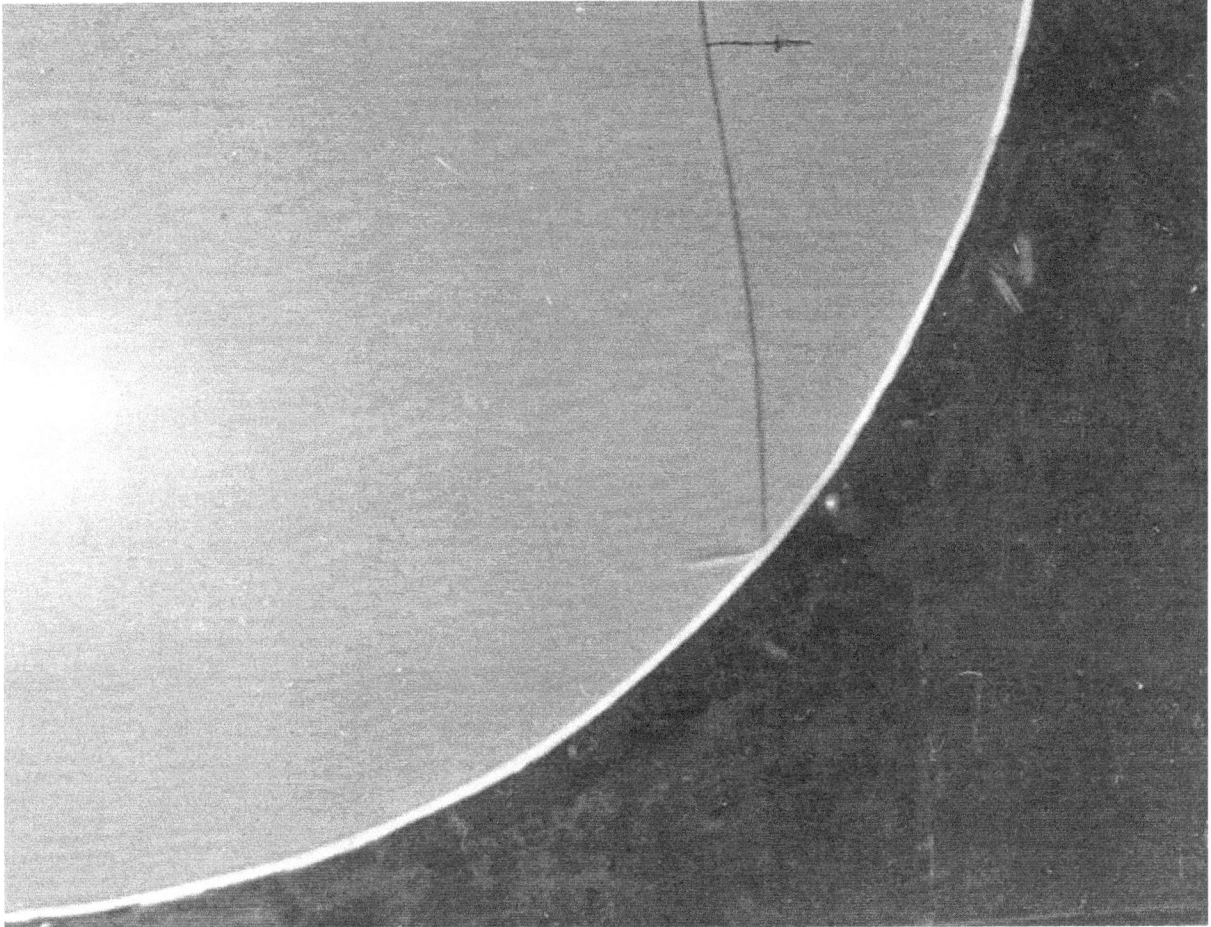

Photo 16
Regular reflection of an incident
shock wave from a concave round surface

a)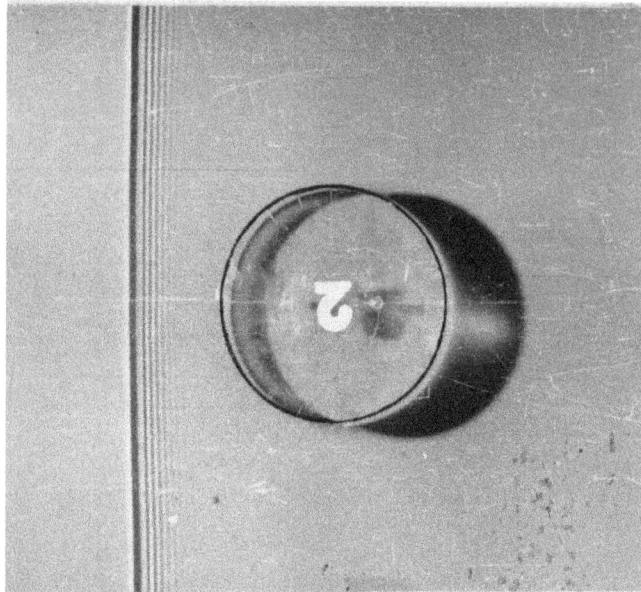

b)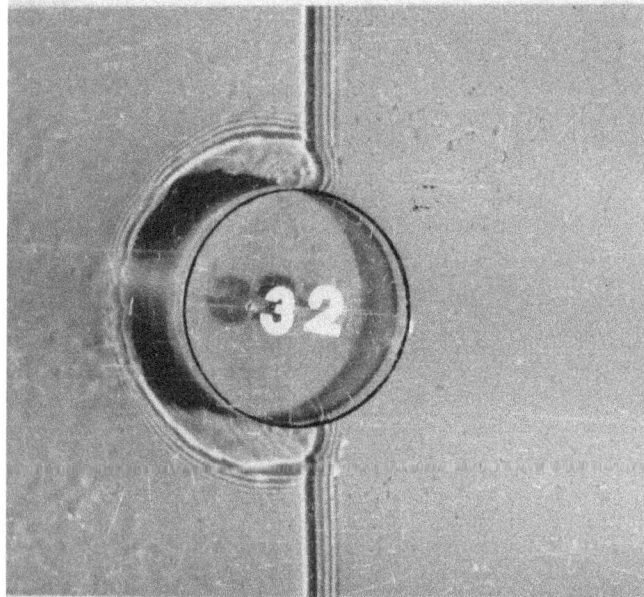

Photo 17.
Irregular reflection of a shock wave moving from left to right.
Two phases are shown:
a. The shock wave has not yet reached the cylinder.
b. The wave has interacted with the cylinder, generating Mach stems and contact discontinuities.

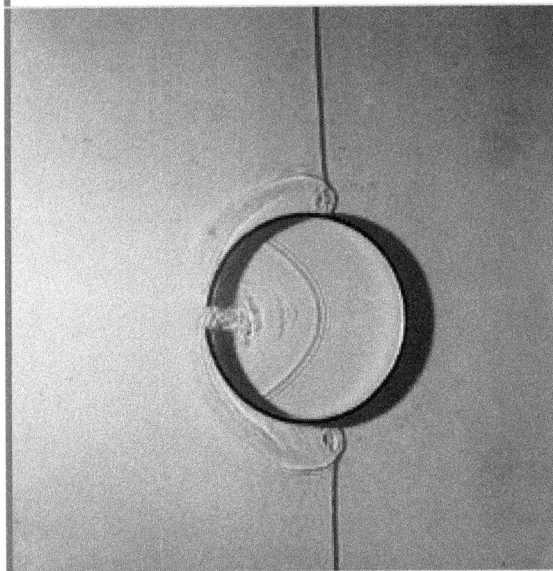

Photo 18, 19, 20
Shock wave interaction with a stationary slotted ring.

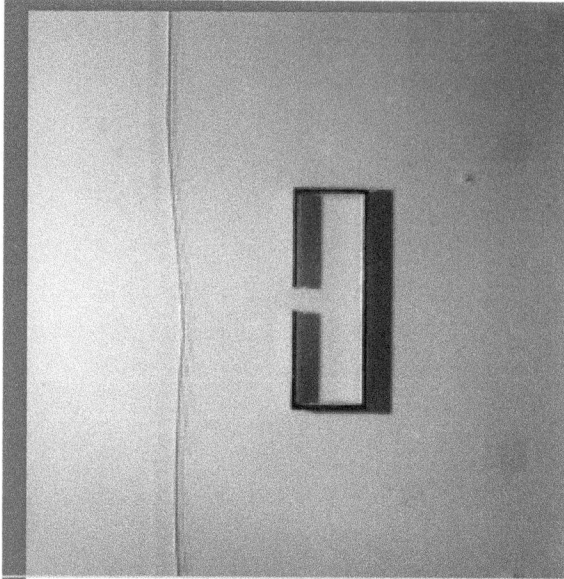

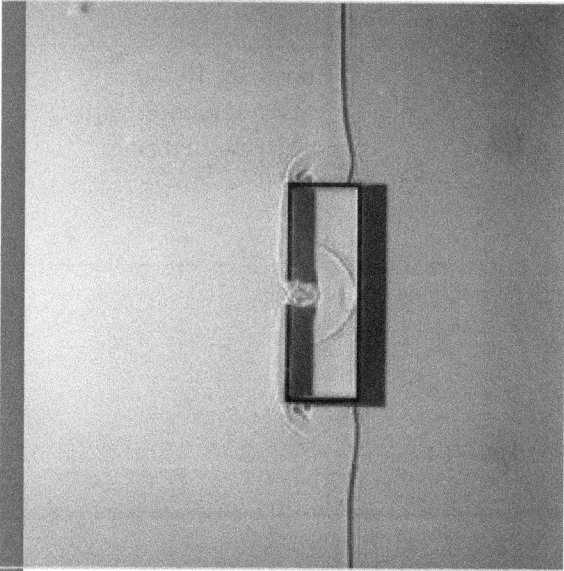

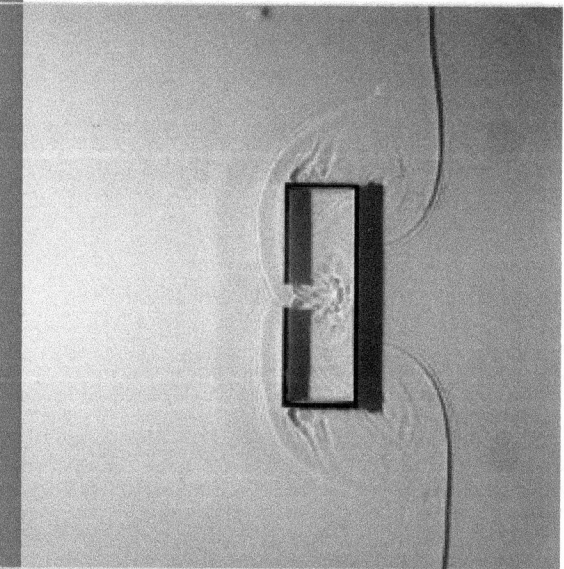

Photo 21, 22, 23

Shock wave interaction with a stationary slotted rectangular chamber.

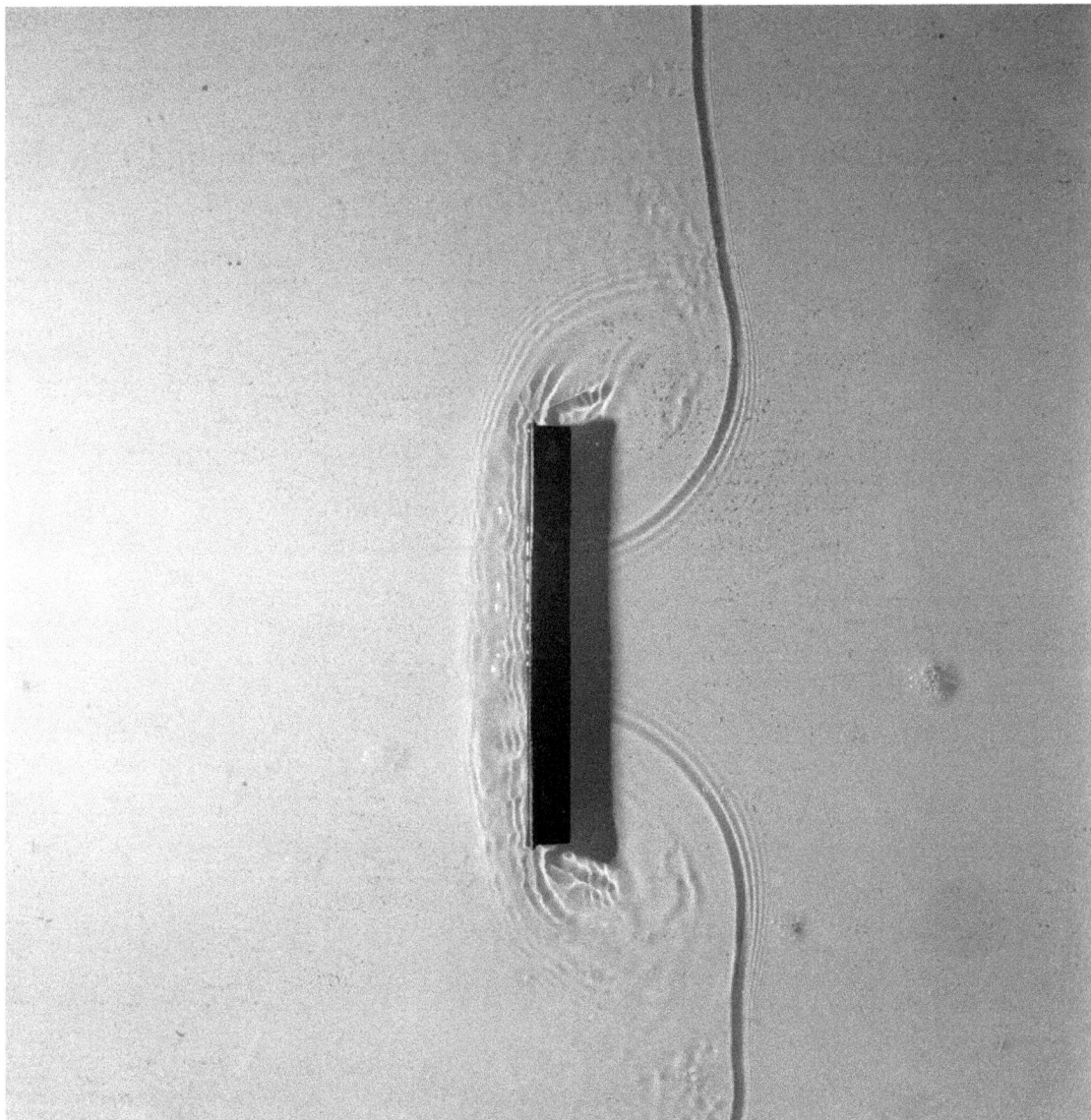

Photo 24

Shock wave interaction with a stationary plate.

Interference and Diffraction of Shock Waves during Flow around Two Separating Bodies

Photos 25, 26.
Visualization of supersonic flow around two
bodies arranged in tandem and moving apart.

The gas-hydraulic analogy (GHA) method is used to

model separation processes based on various

scenarios: deceleration of the rear body, movement

along divergent trajectories, acceleration of the front

body after detachment, and others.

Photo 25

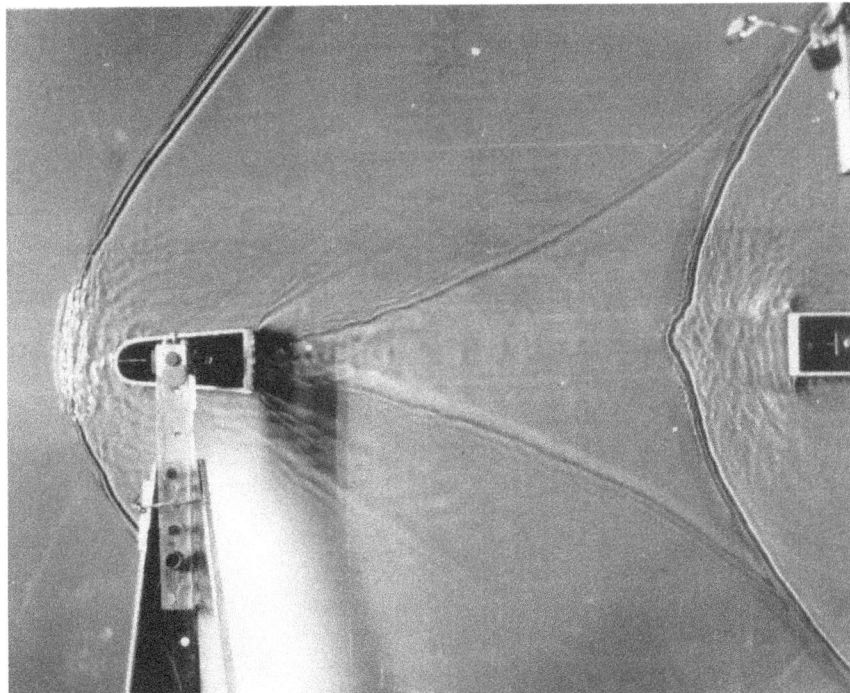

Photo 26

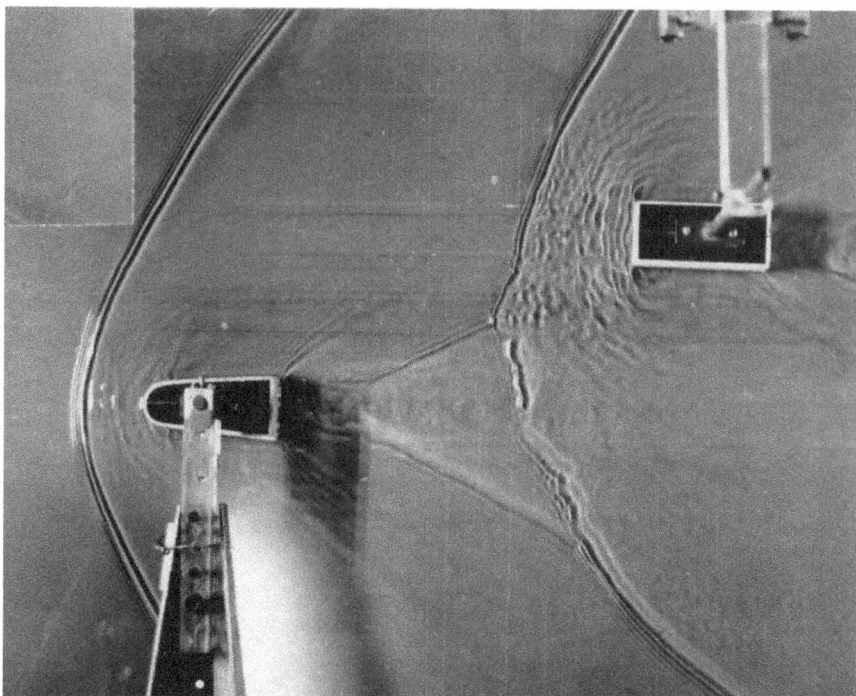

Photo 27

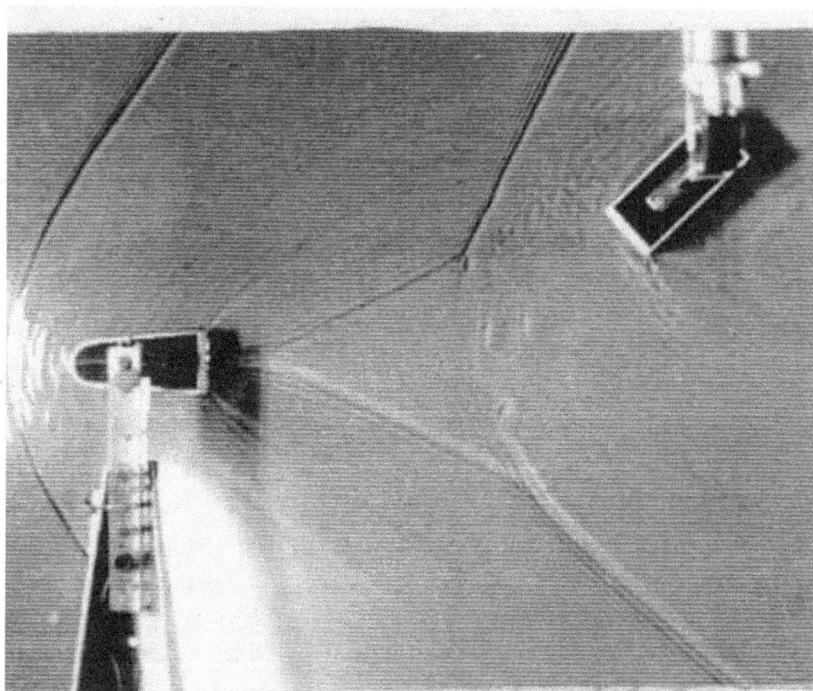

Photo 28

Photo 29

Photo 30

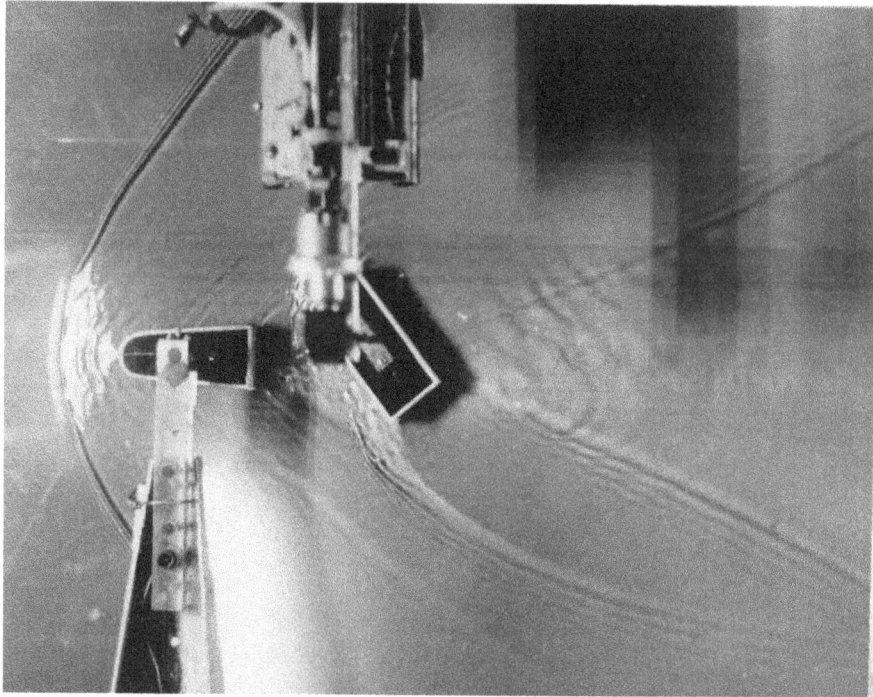

Photo 31

Photos 27, 28, 29, 30, 31.

Shadow images. illustrating various scenarios of rear body rotation during translational motion.

The Gas-Hydraulic Analogy method is employed to simulate dynamic separation phenomena in supersonic flow environments by modeling a variety of engineering scenarios: rear-body deceleration during staged separation, trajectory deflection of the trailing component, post-separation acceleration of the leading body, and other kinematic configurations. These simulations capture key flow phenomena such as shock wave interactions, expansion regions, and unsteady wake structures—providing valuable insights for the design and analysis of high-speed aerospace systems.

Investigation of Shock Wave Structures in the Flow around Flow-Splitting Bodies

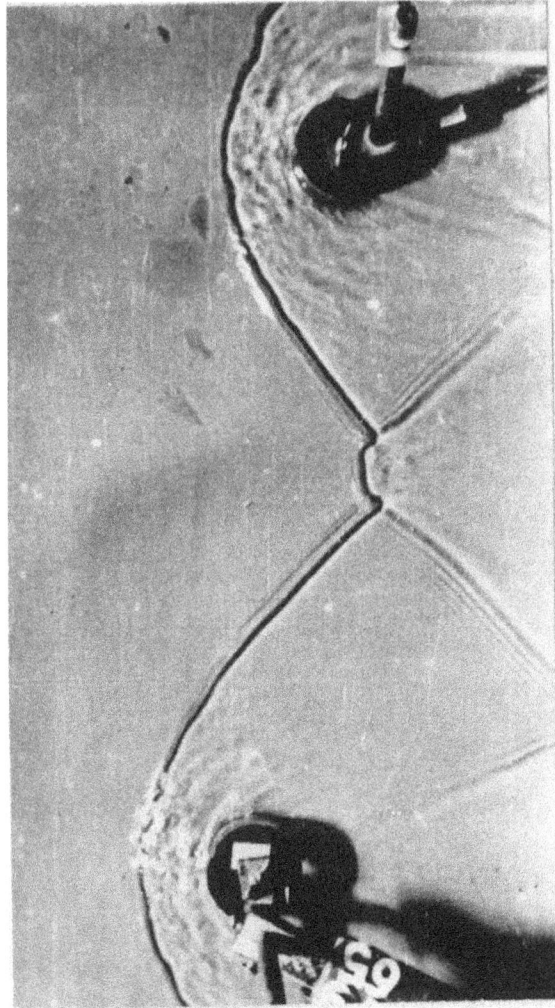

Photo 32
Triple shock wave configuration with a
tangential discontinuity extending from the point of
intersection with a turbulent zone — a phenomenon
known as Mach reflection. Mach number $M_\infty = 1.9$

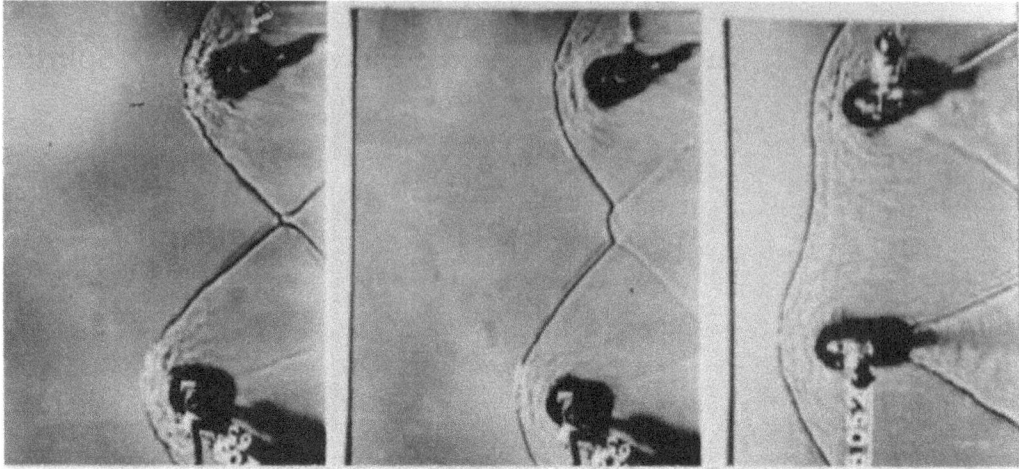

Photo 33
Variation in flow behavior with different Mach number

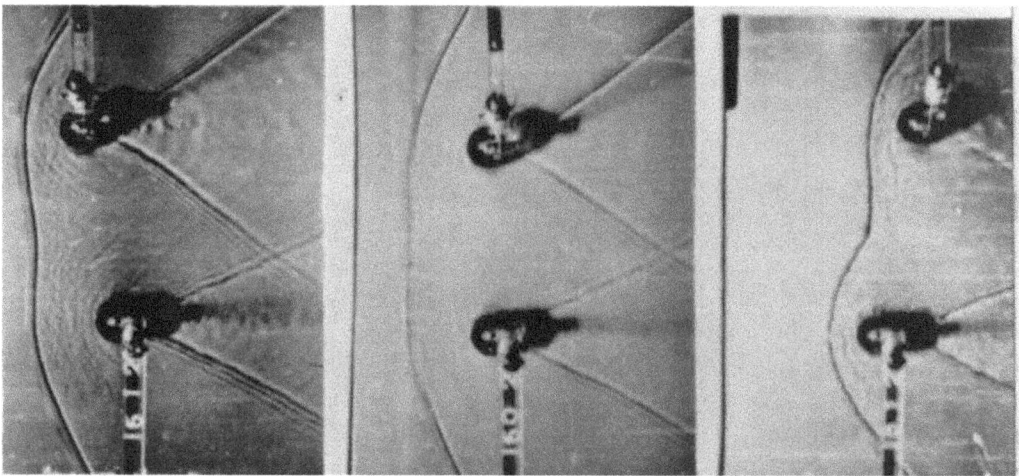

Photo 34
Variation of flow behavior with the position of the lower body relative to the
upper body along the x-axis of motion.

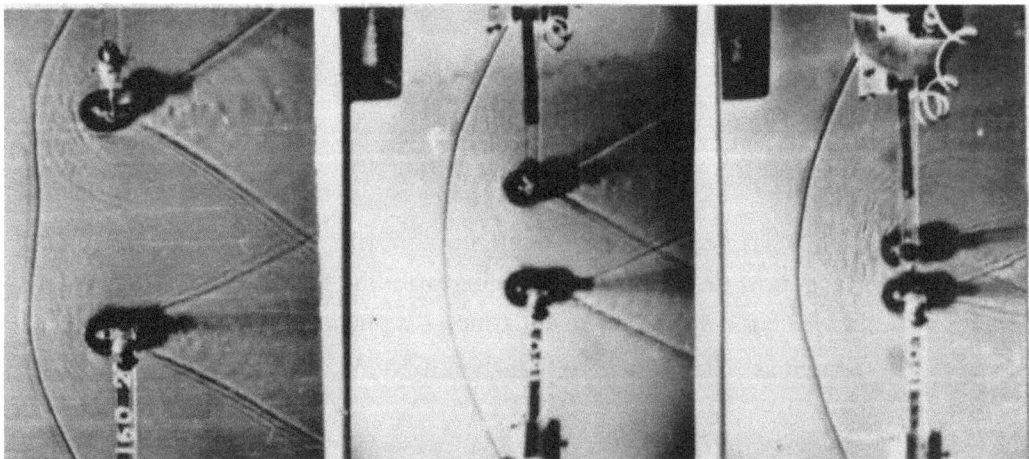

Photo 35
Motion of bodies during mutual displacement along the y-axis.

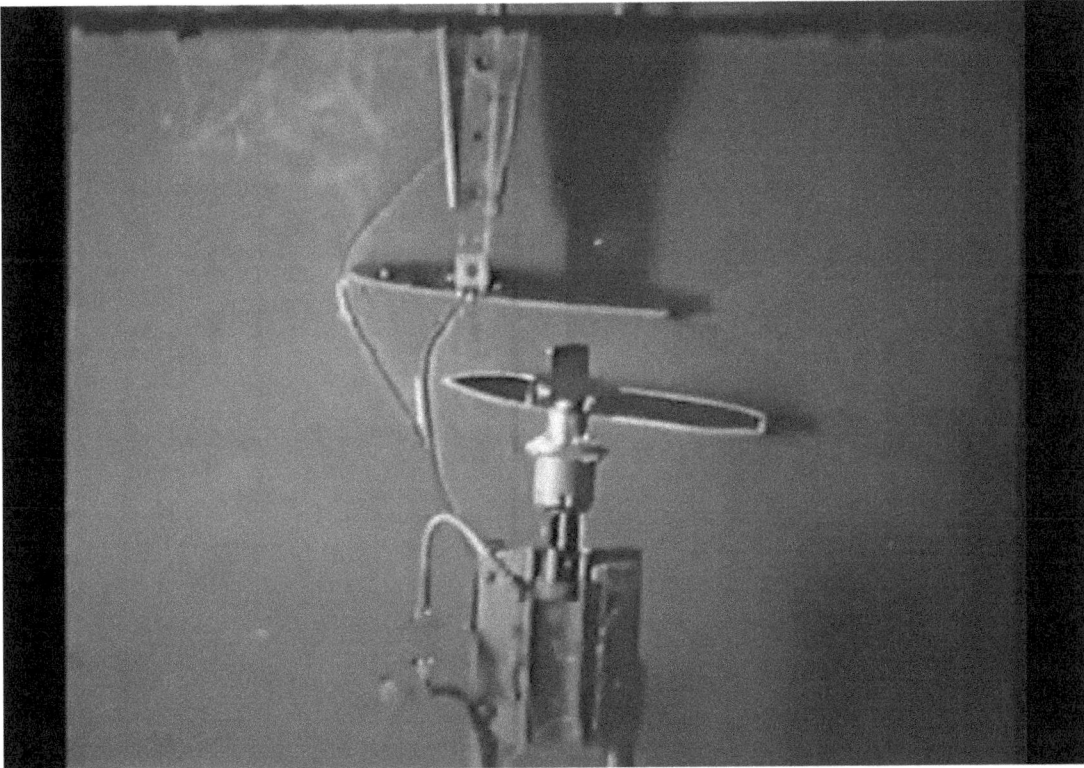

Photo 36

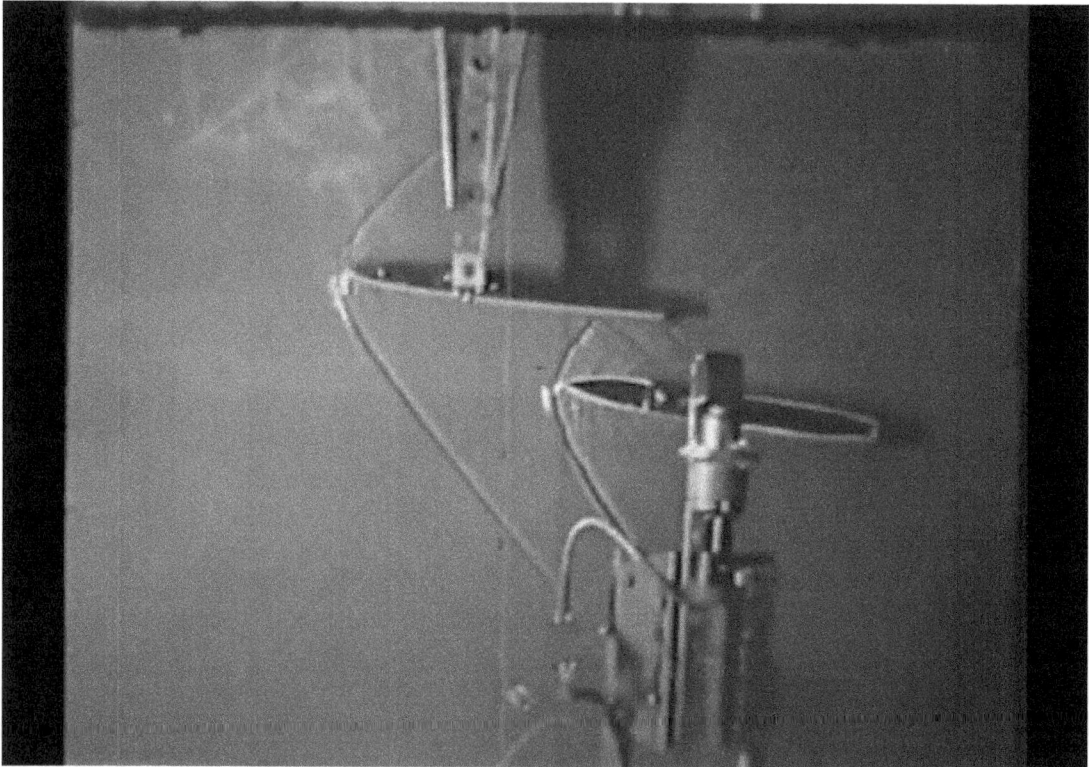

Photo 37

Photo 38

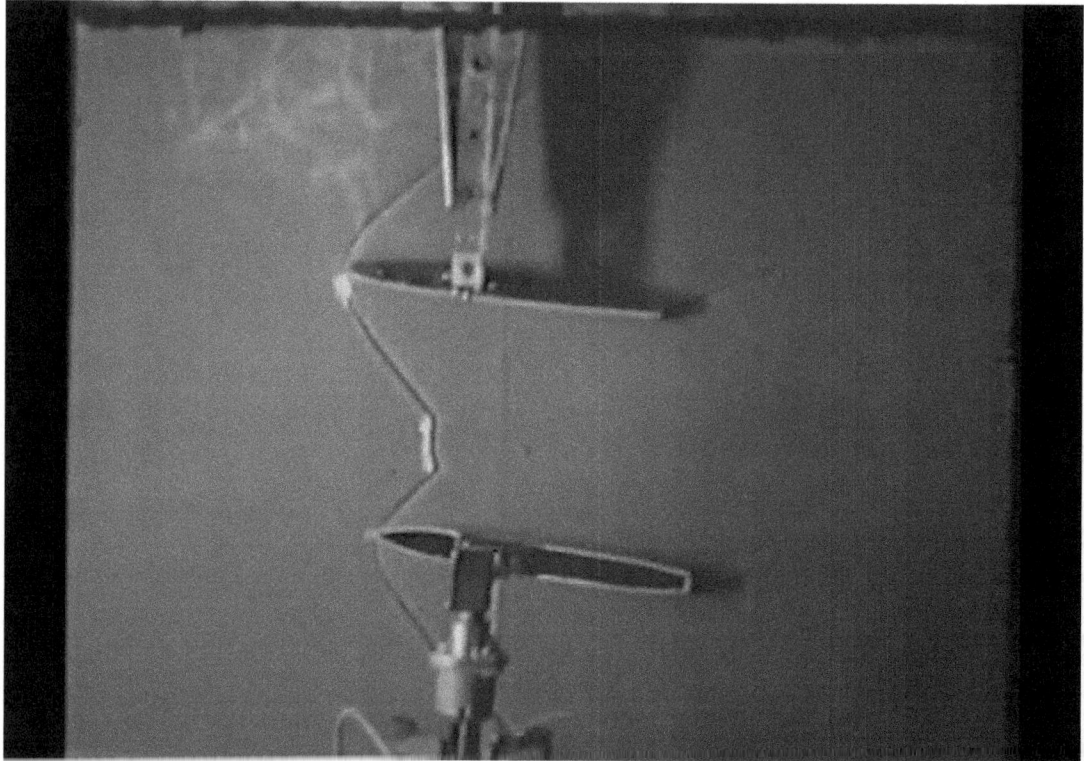

Photo 39

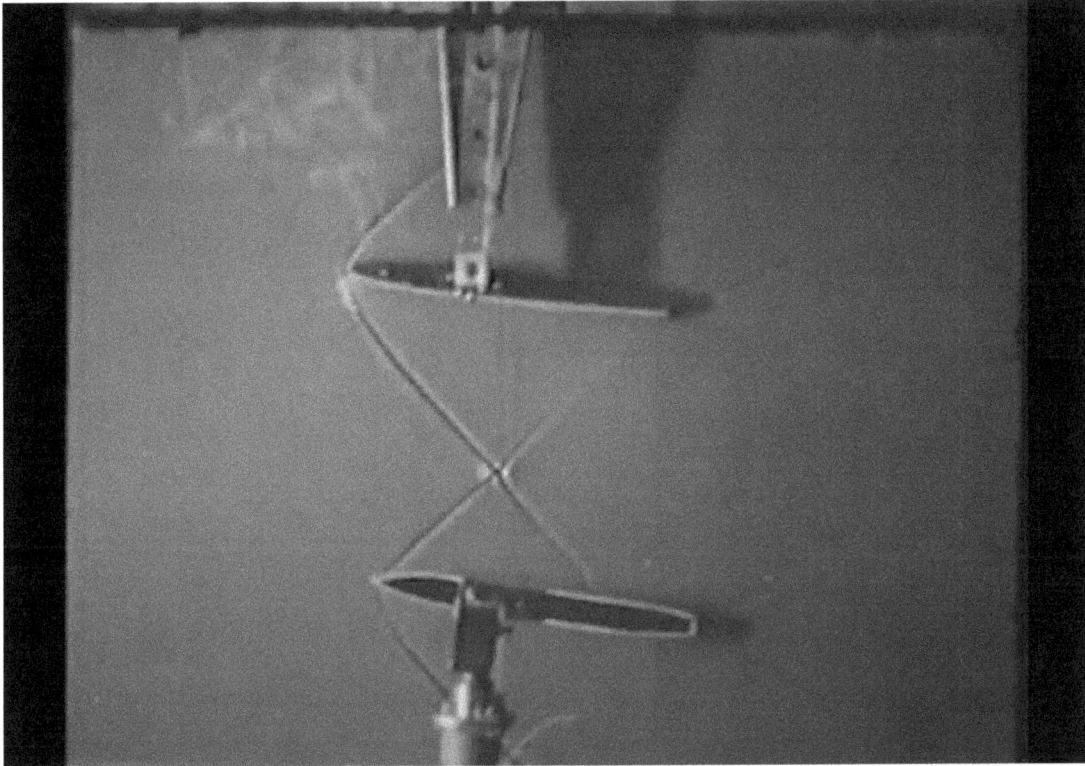

Photo 40

Photos 34, 35, 36, 37, 38

Individual frames from the film depict the flow around two separating airfoils.

The studies were conducted under the following parameters:

- **Mach number range:** 1.2 to 2.5
- **Lower airfoil axial displacement:** -0.5 to 1.0 *(relative to the chord length of the upper airfoil)*
- **Lower airfoil movement range:** 0 to 1 *(relative to the chord length of the upper airfoil)*
- **Upper airfoil angle of attack:** 0° to 15°
- **Lower airfoil angle of attack:** -15° to +15°

Tests were performed at varying relative speeds between the lower and upper airfoil.

Note: All linear dimensions are normalized by the chord length of the upper airfoil.

Photo 41

Flow past wedge profiles at $M_\infty = 2.2$; $wedge\ angle\ \theta = 12^0$ ($flow\ deflection$), shock angle $\beta = 38^0$ (oblique shock inclination).

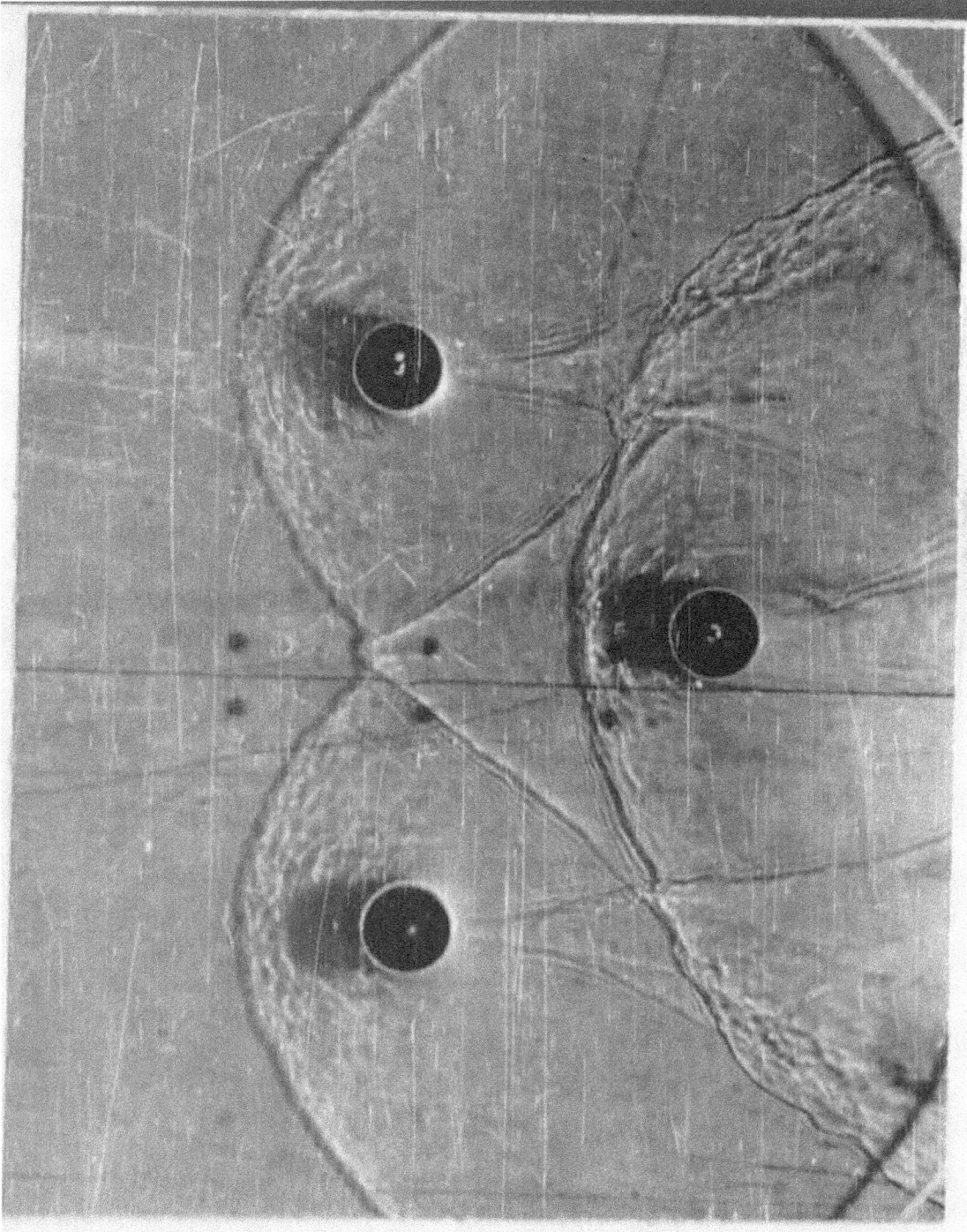

Photo 42

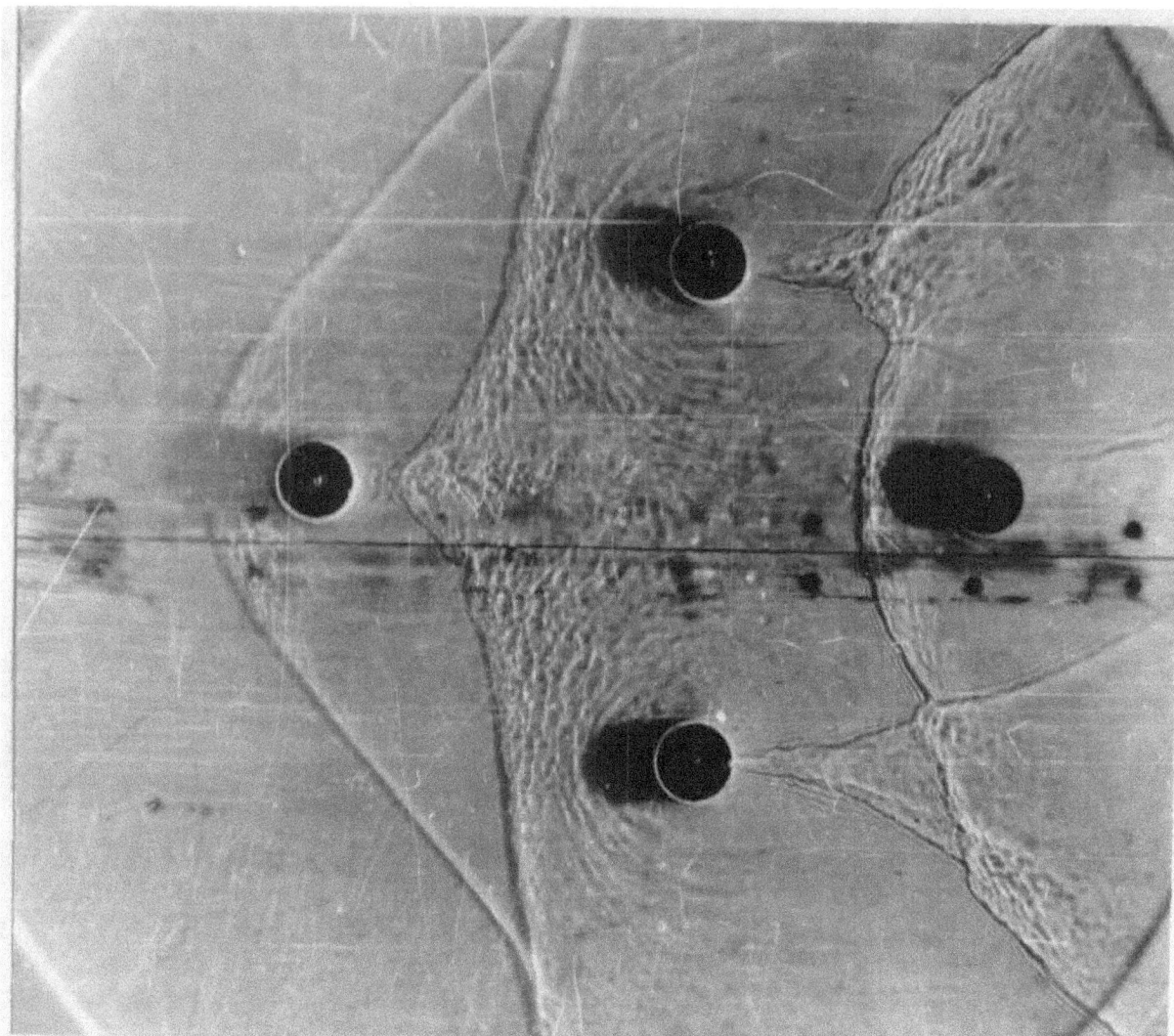

Photo 43

Photo 42, 43, 44
Hydraulic analogy
modeling of the separation
(or joining) process of
bodies at supersonic speed.

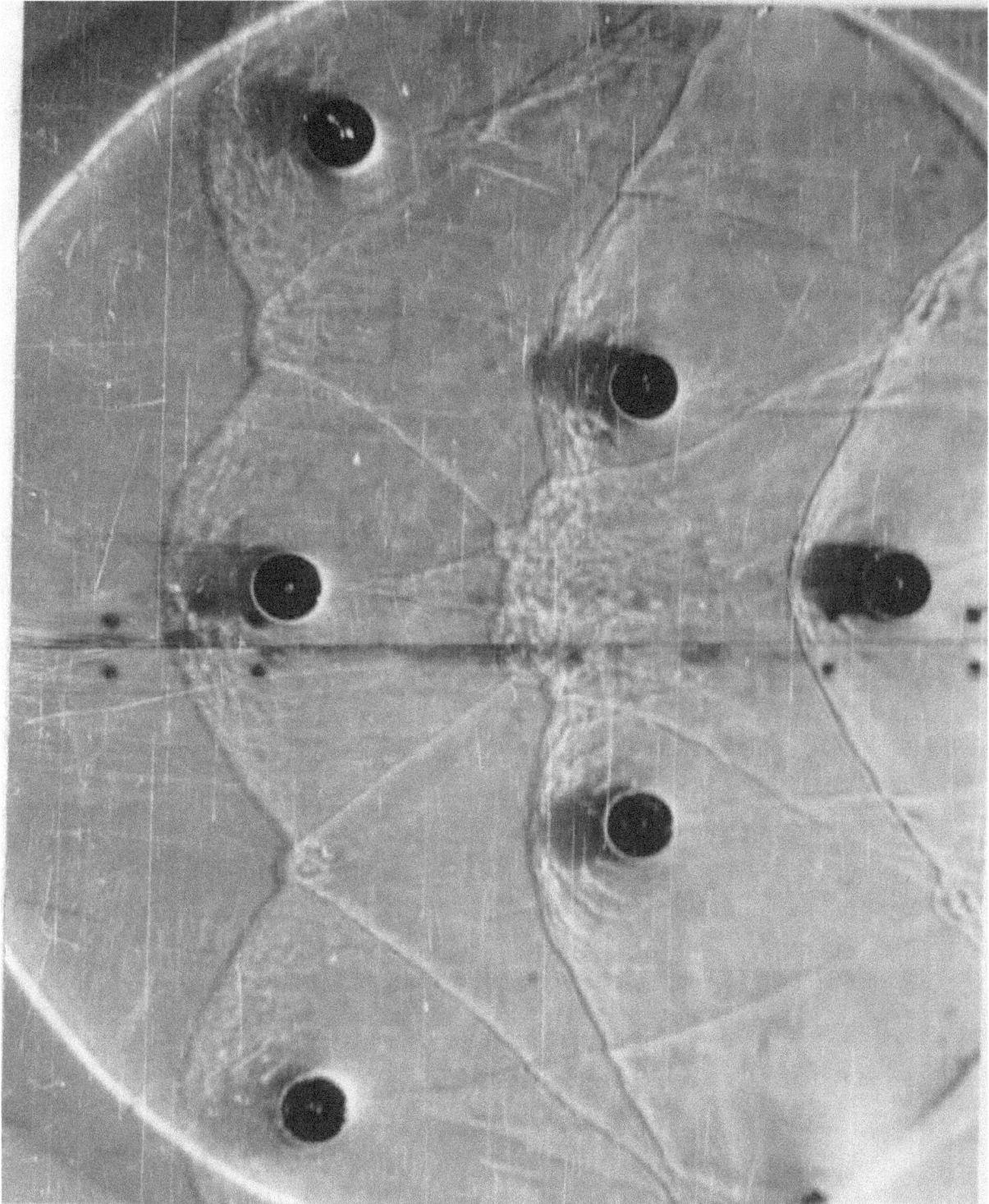

Photo 44
At a certain moment, a single body moving at supersonic speed breaks into smaller bodies, which can move indifferent directions at varying speeds

Photo 45
Wedge-shaped profile in free flight, $M_\infty = 1.5$.

Photo 46

Photo 47

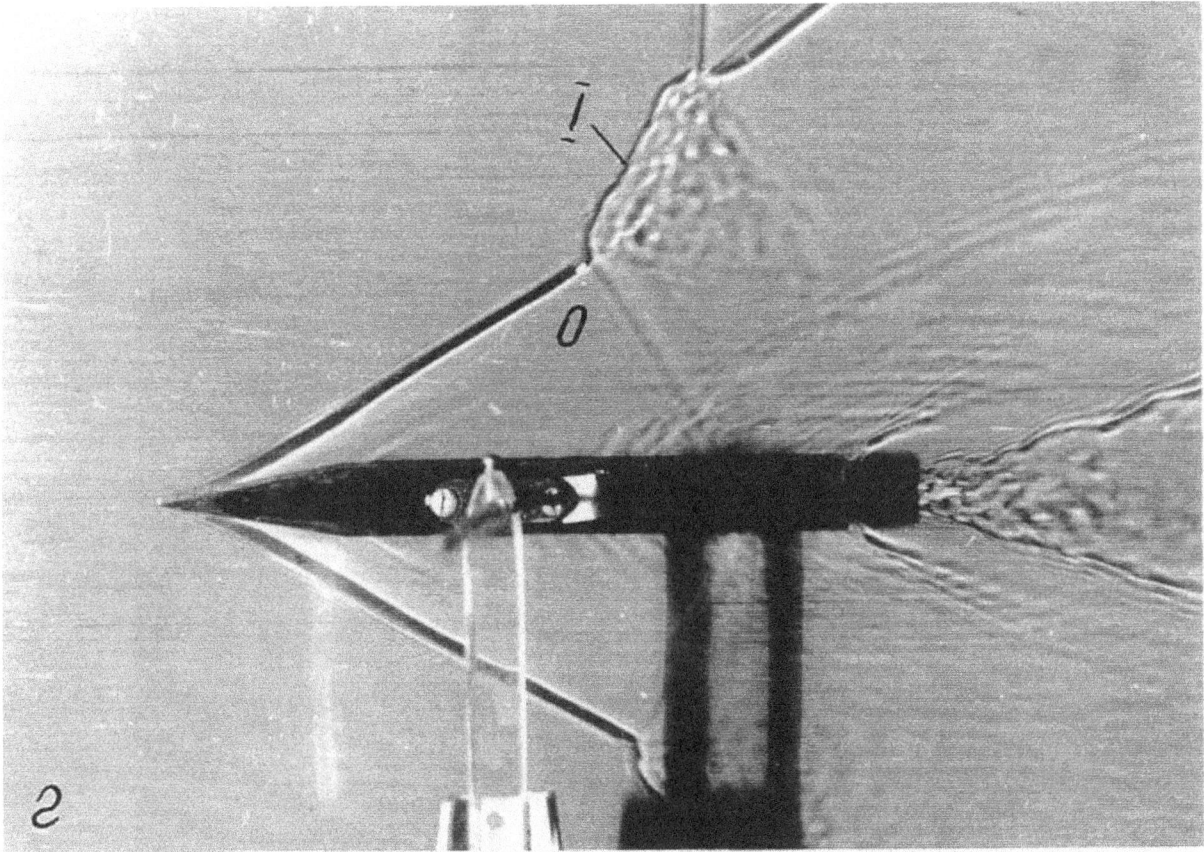

Photo 48

Photos 45 Wedge-shaped profile in free flight, M_∞=1.5. **Photo 46, 47, 48** Overtaking interaction of shock waves. Three phases.

A wedge-shaped profile with an attached shock wave (Mach number = 3.7)

overtakes and interacts with the front of another shock wave (Mach number = 1.5) moving in the same direction (from right to left)

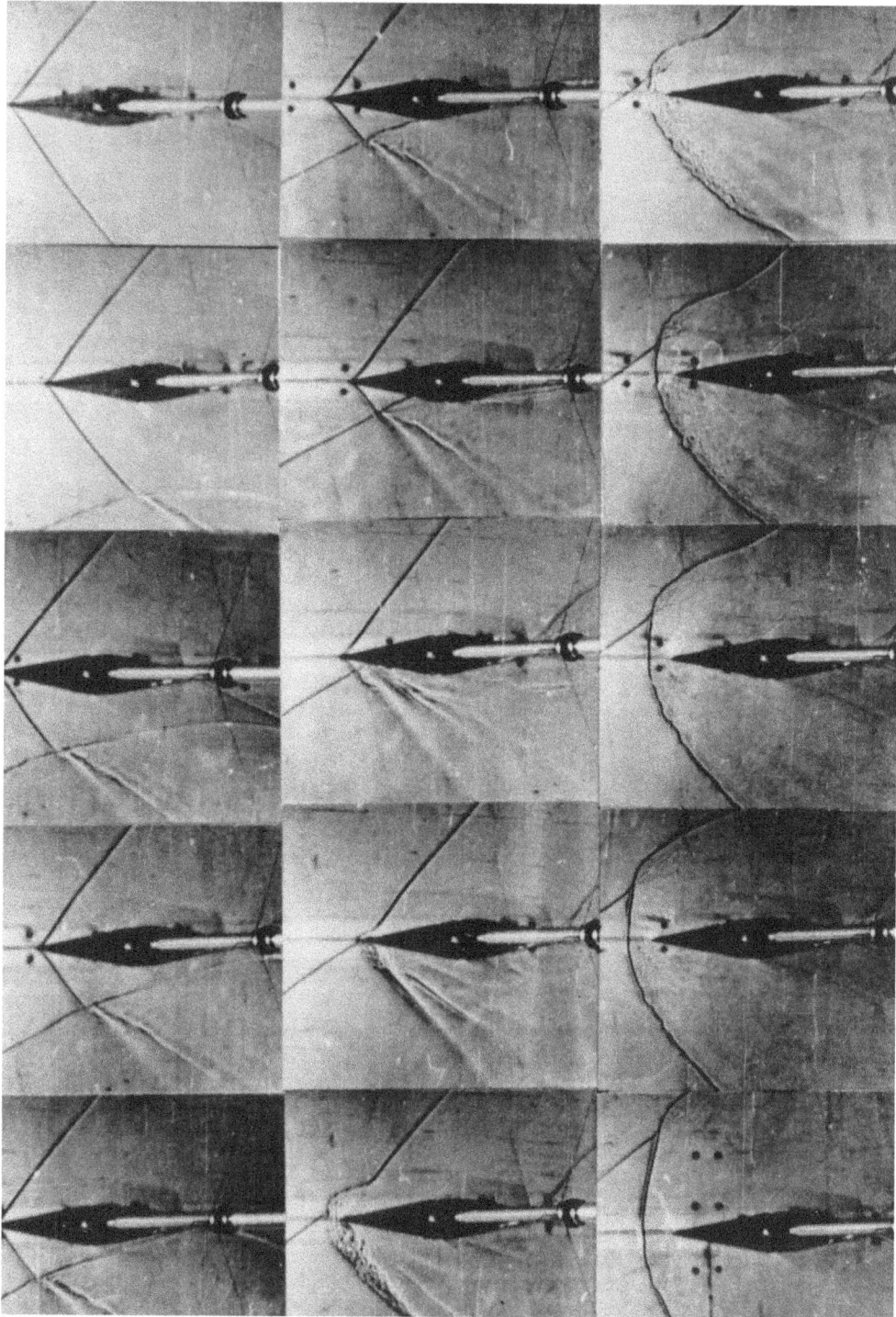

Photo 49

The cinematogram illustrating the interaction of an inclined overtaking shock wave (incidence angle: 115°, shock Mach number: 1.8) with a body moving at Mach 1.5.

Hydraulic analogy modeling allows the study of the impact of single or multiple shock waves,

or moving bodies, at any angle of wave incidence over the full range from 0 to 360 degrees.

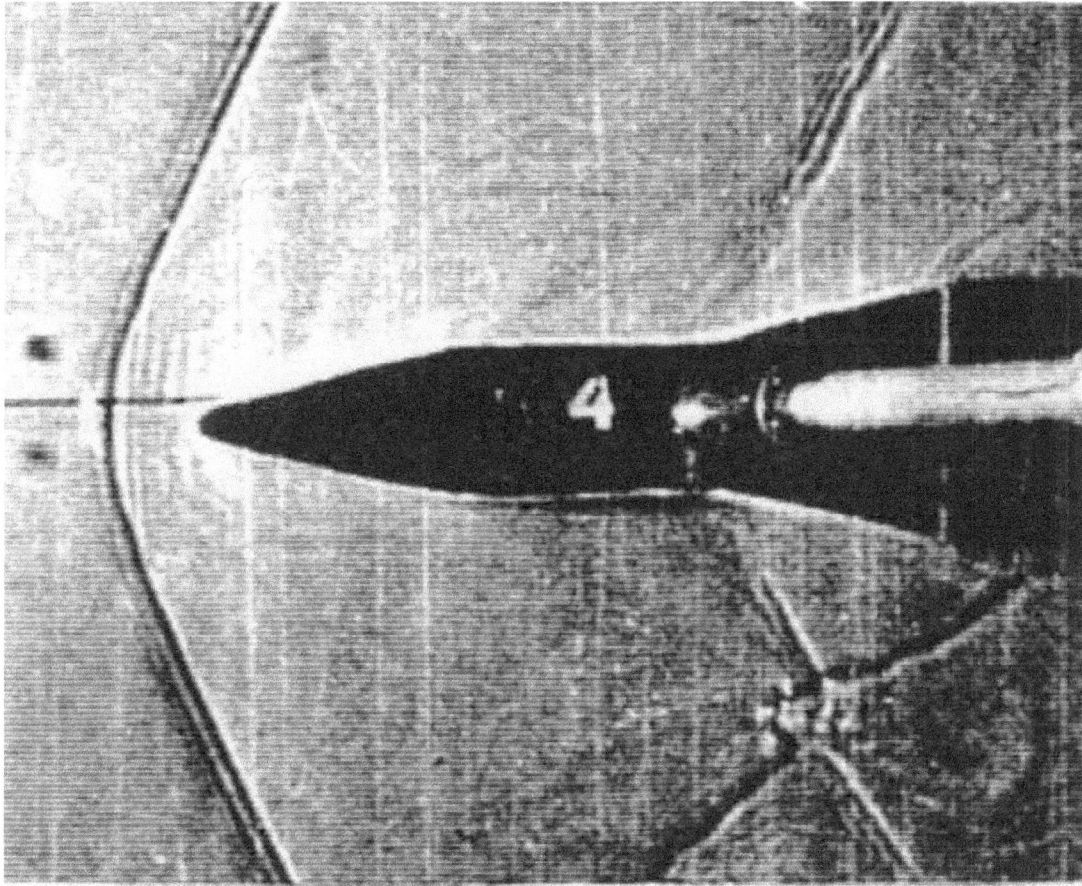

Photo 50
Interaction of shock waves from a profile
moving at supersonic speed with an incoming
shock wave from below.

Photo 51
The shock wave approaches from behind,
perpendicular to the x-axis of motion, overtakes the
body, and interacts with the leading shock wave.

Photo 52

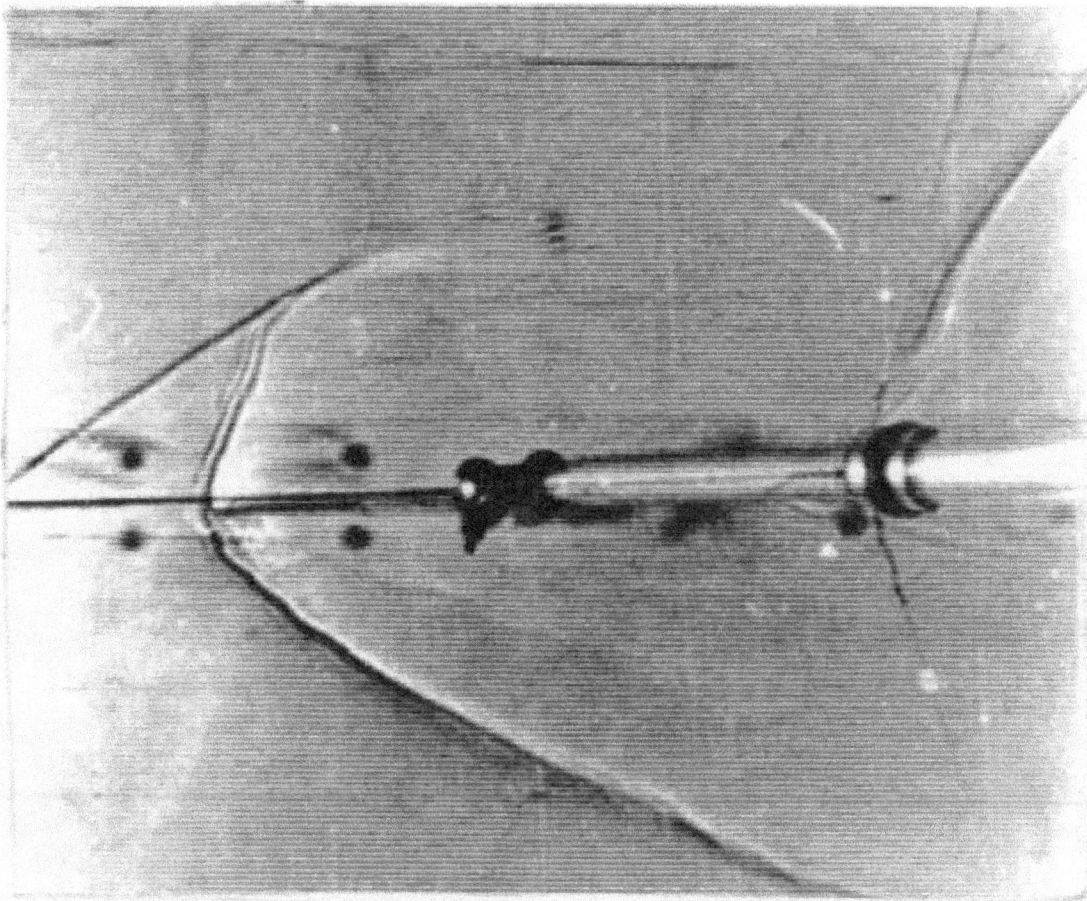

Photo 53
Photo 52, 53 The shock wave approaches from behind and below at an angle to the axis of motion, overtakes the body, and interacts with the leading shock wave.

Shock wave influences back from bottom at an angle related to the axis of movement. Catches the bode and interact with the leading shock wave.

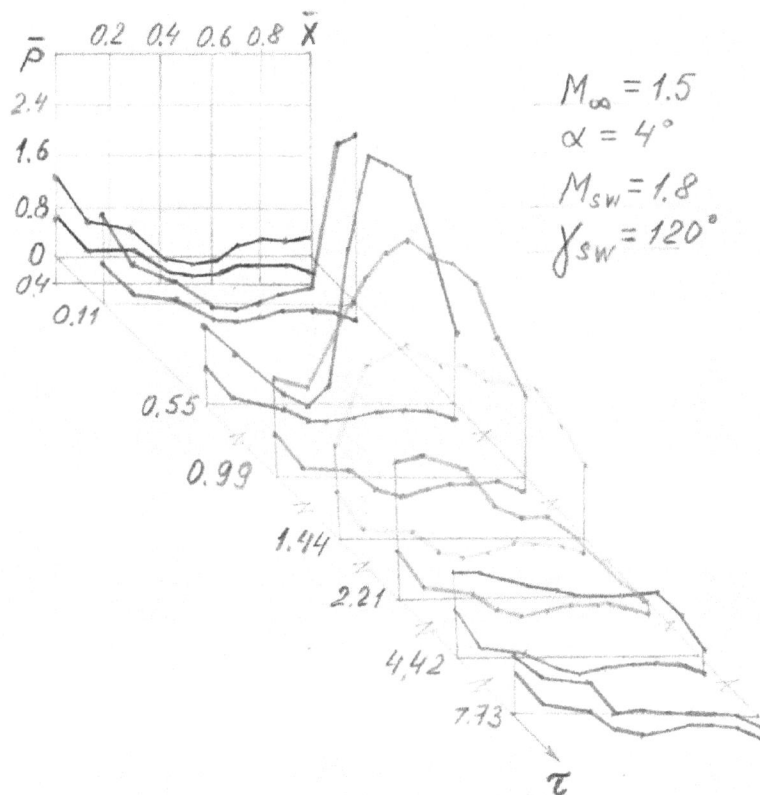

Photo 54

Pressure distribution diagrams along the profile at M∞ = 1.5, α = 4°, M$_s$w = 1.8, γ = 120° (solid line – upper surface; dashed line – lower surface).
The image presents pressure distribution diagrams for a flat-plate profile moving at supersonic speed, illustrating the influence of interaction with an overtaking shock wave at various time intervals.

Development of shock wave structures during the simultaneous impact of two shock waves on a stationary cylinder

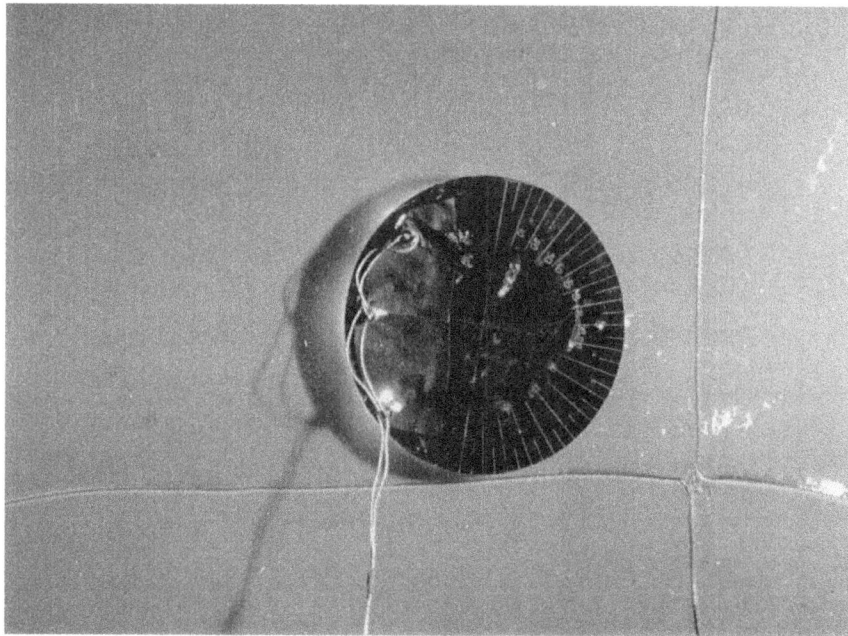

Photo 55
Interaction of two shock waves upward from below, the other from right to left.

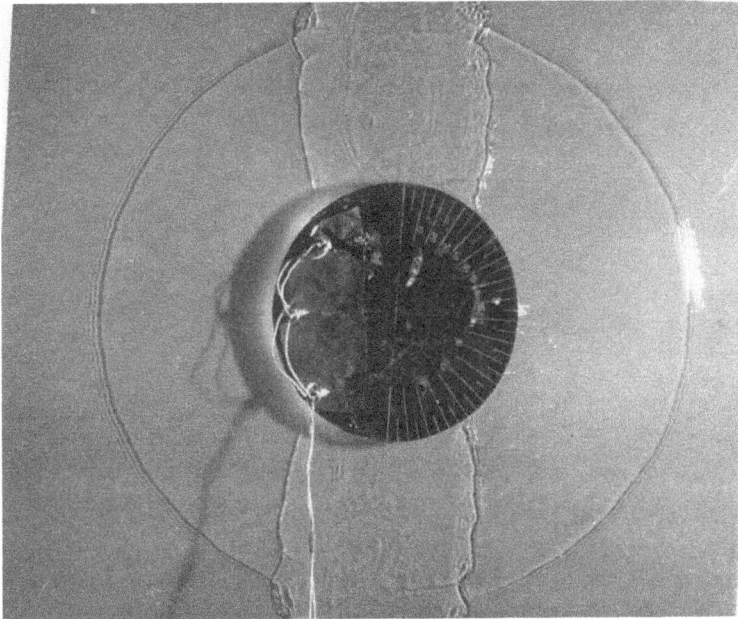

Photo 56

Interaction of shock waves: from left to right – from right to left.

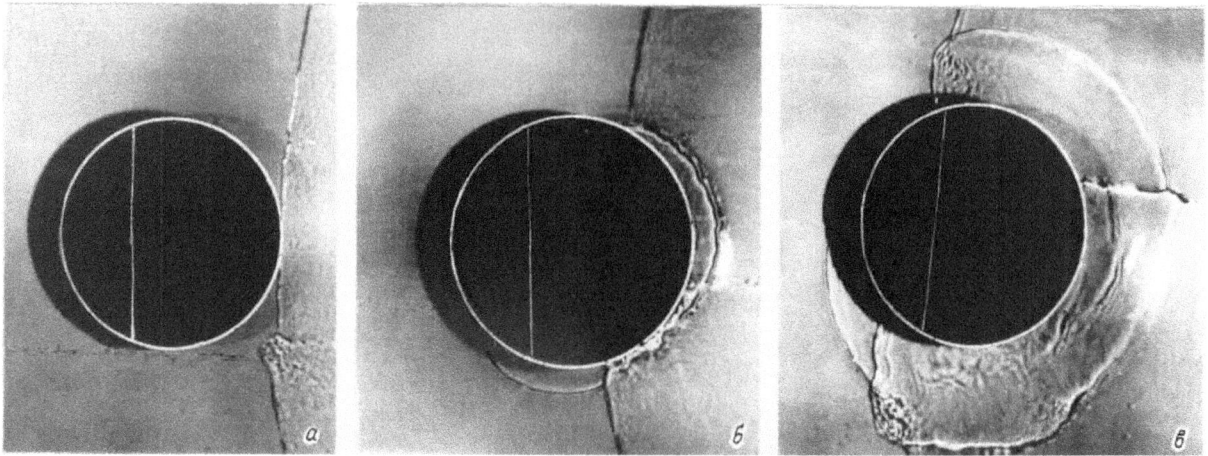

Photo 57
Three stages of interaction between two opposing shock waves—one originating from below and the other from the right—with a stationary cylinder.

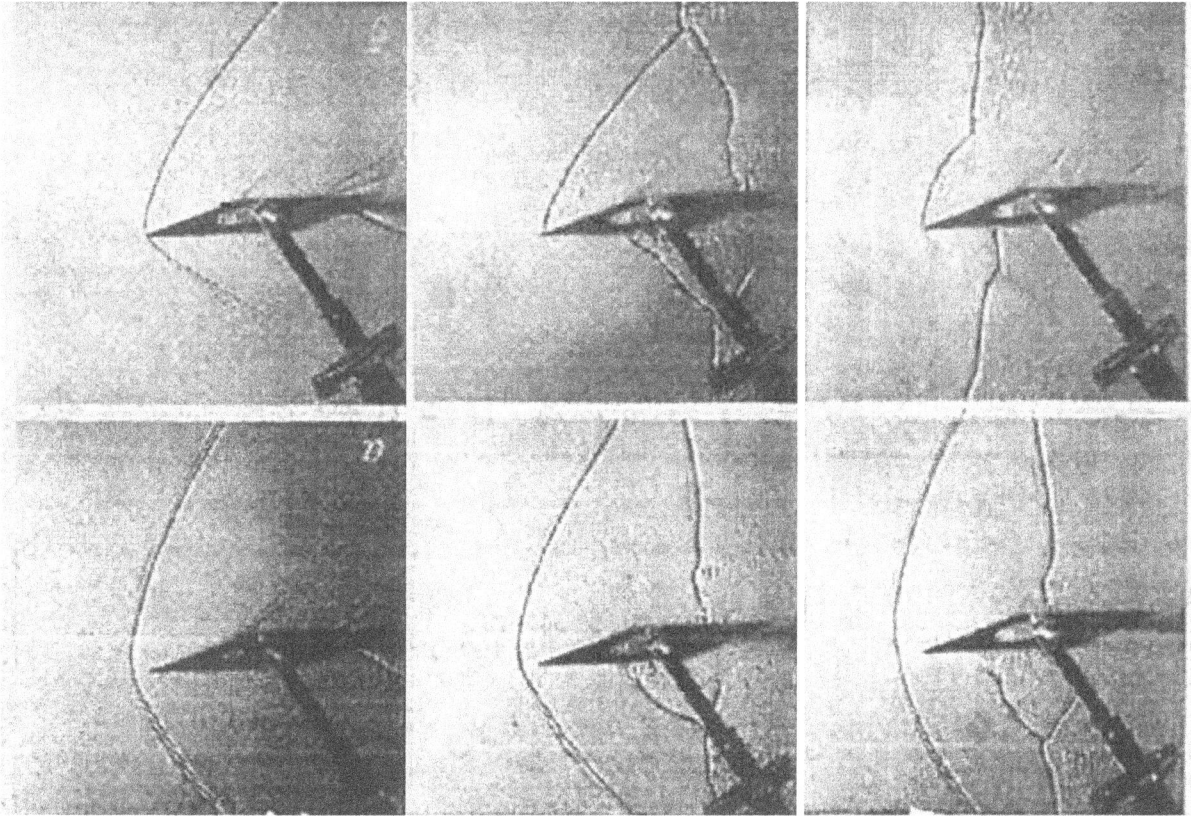

Photo 58

Simulation of the interaction between a supersonically moving body and an overtaking shock wave.

Photo 59

Sequential interaction of two shock waves with a fixed slotted ring. In the image, both shock waves propagate from left to right.

Flow around an Airfoil in the Vicinity of a Solid Screen of Various Shapes

Gas Hydraulic analogy simulation of supersonic flow past a moving body near a non-uniform surface boundary.

The photos 60, 61, 62, 63, 64, 65, 66, 67, 68 capture segments of body motion near a surface featuring local elevations, depressions, and a wavy profile (Mach number $M_\infty = 1.6$–1.7).

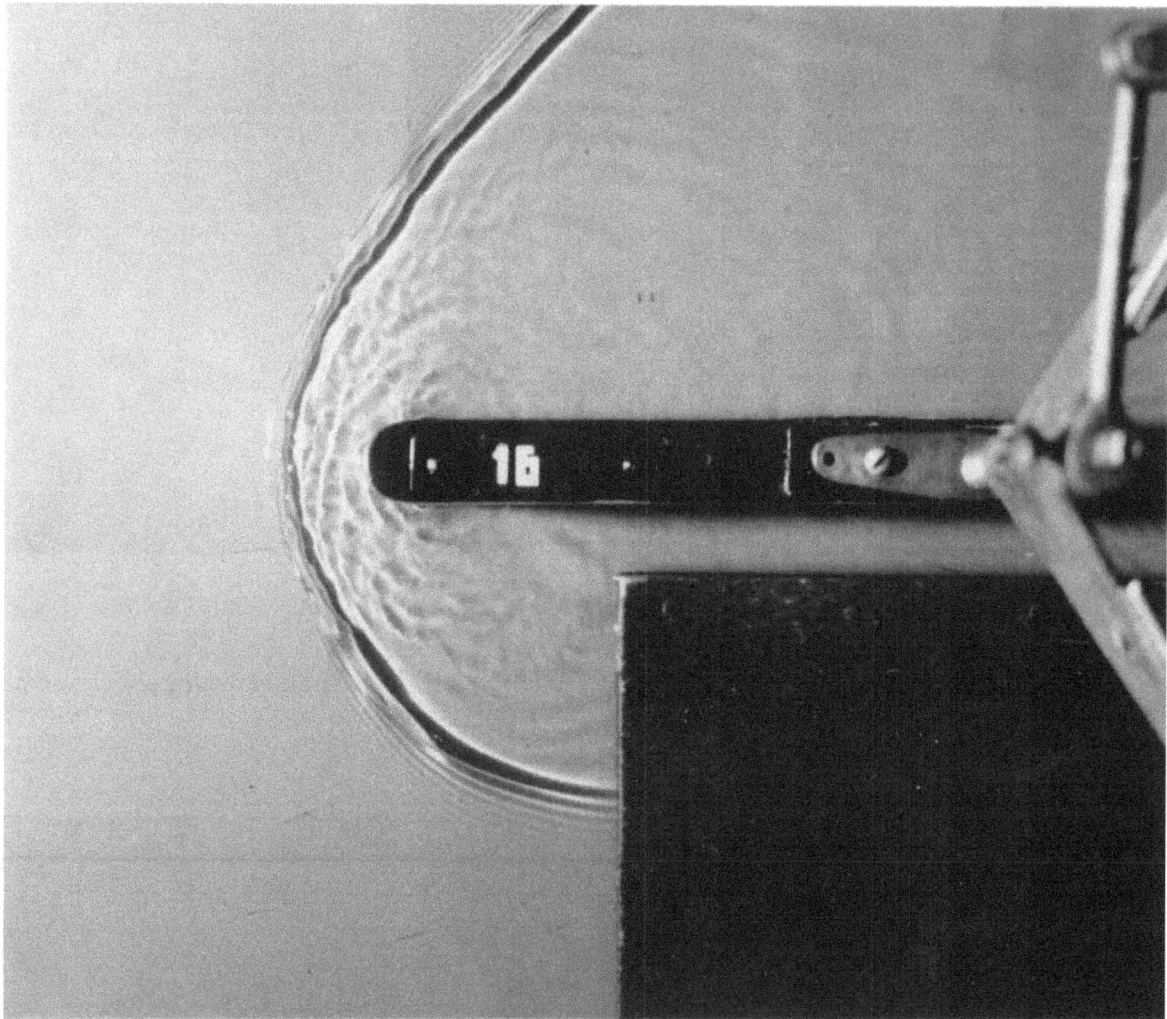

Photo 60

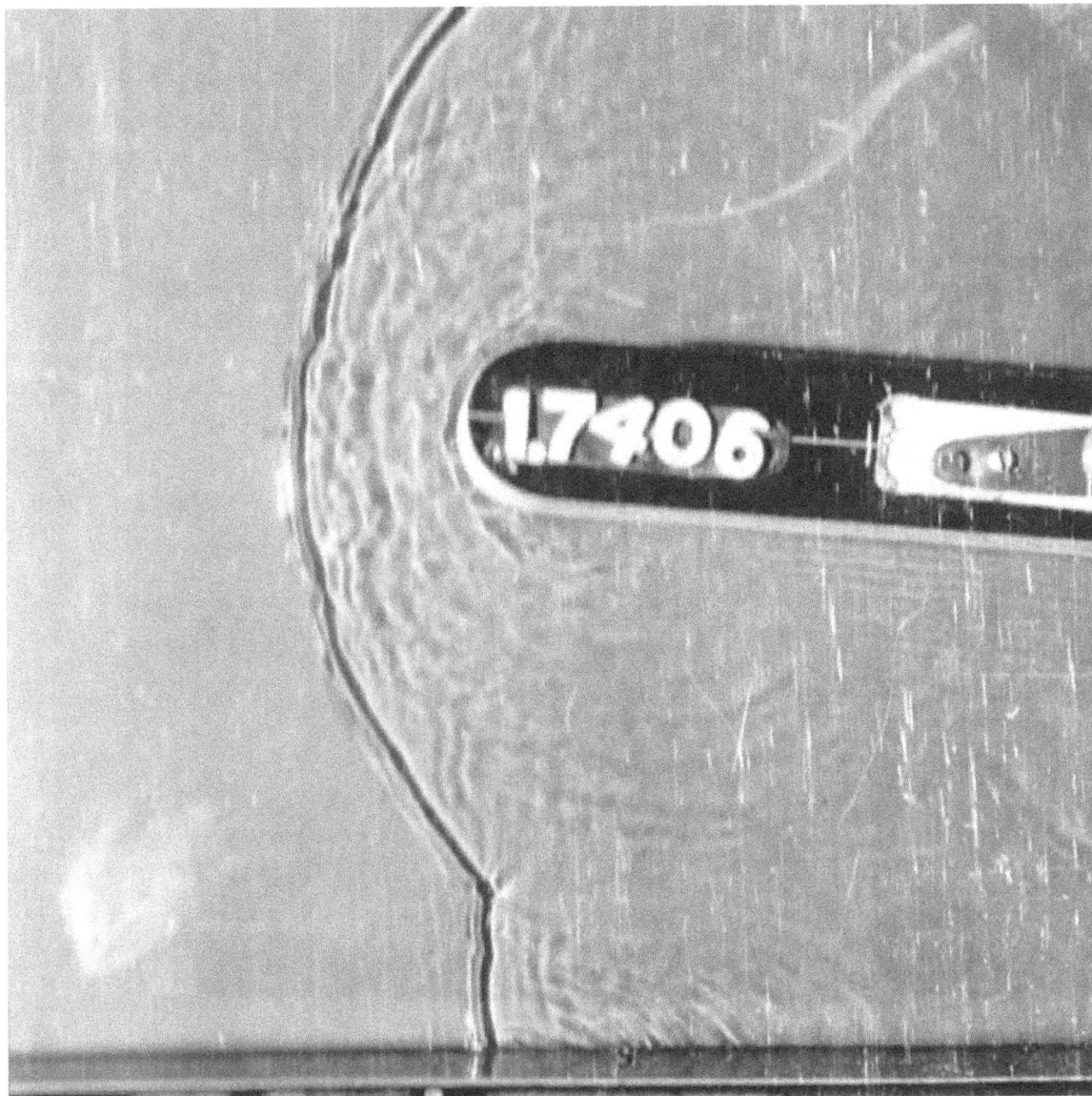

Photo 61

The photo 61 captures supersonic flow ($M_\infty = 1.7$) around a rounded-nose profile near a flat screen. The dimensionless gap between the profile and the screen is 0.6 body lengths, with an angle of attack $\alpha = 4°$. Irregular reflection of the bow shock is observed.

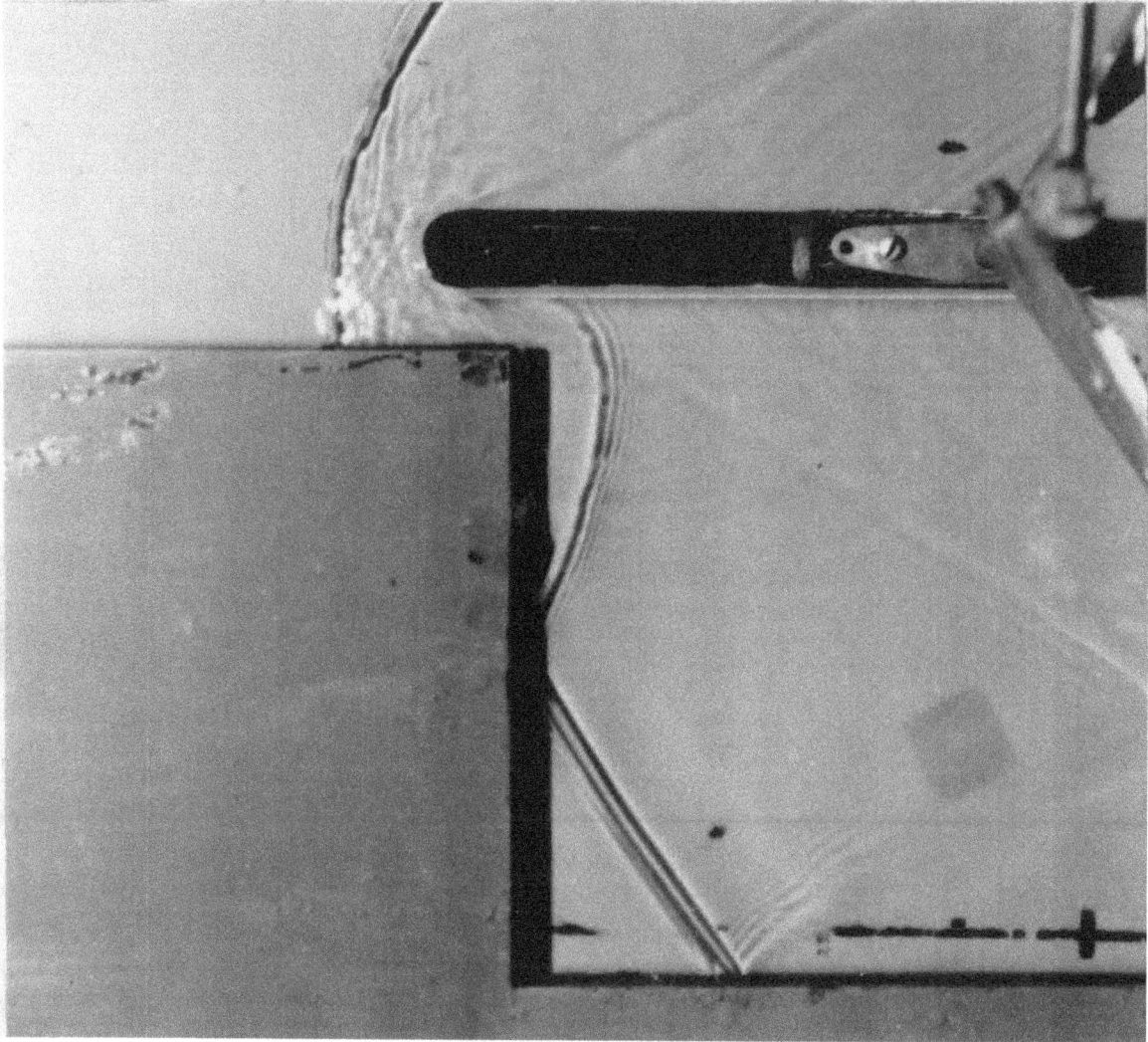

Photo 62

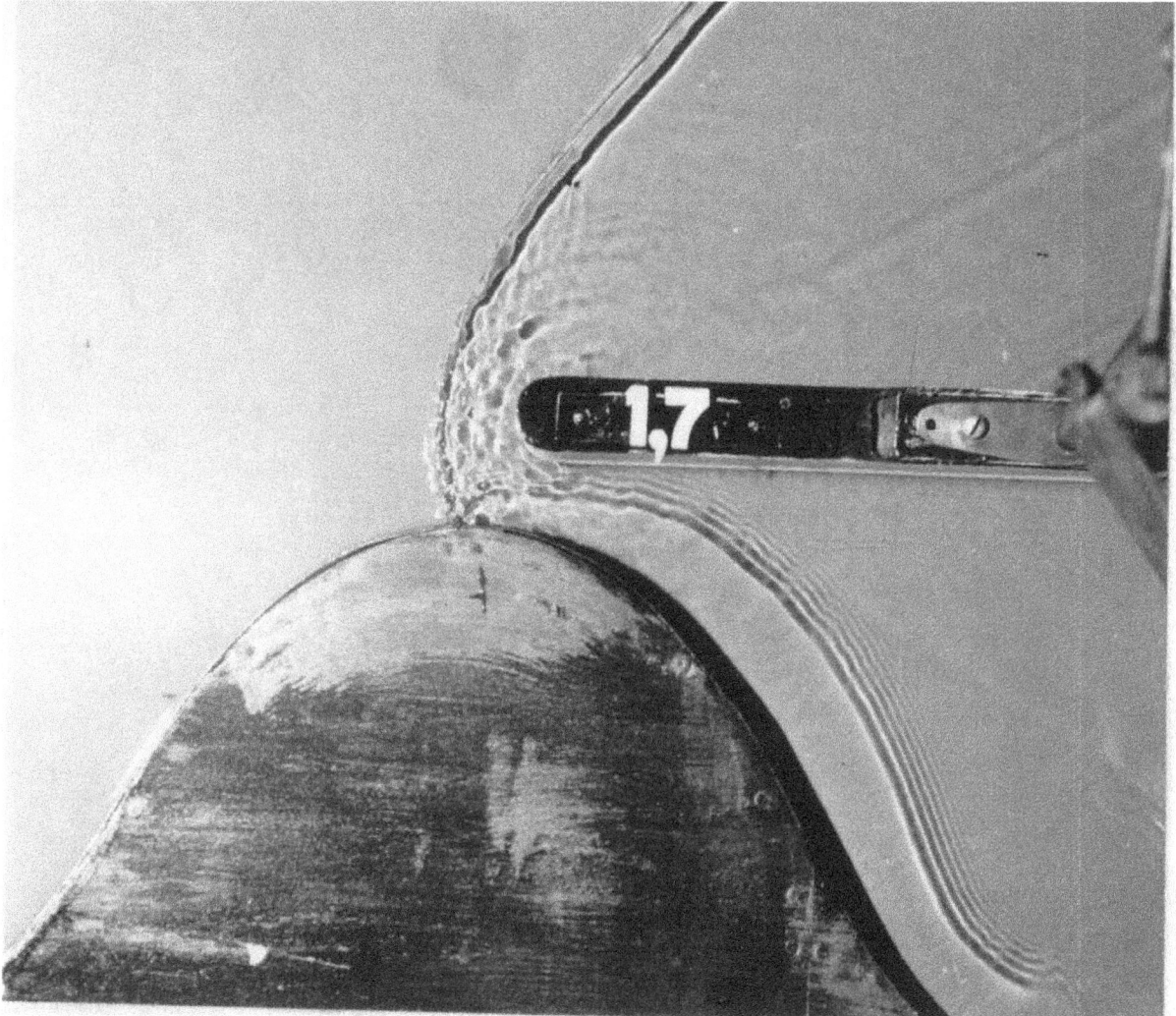

Photo 63

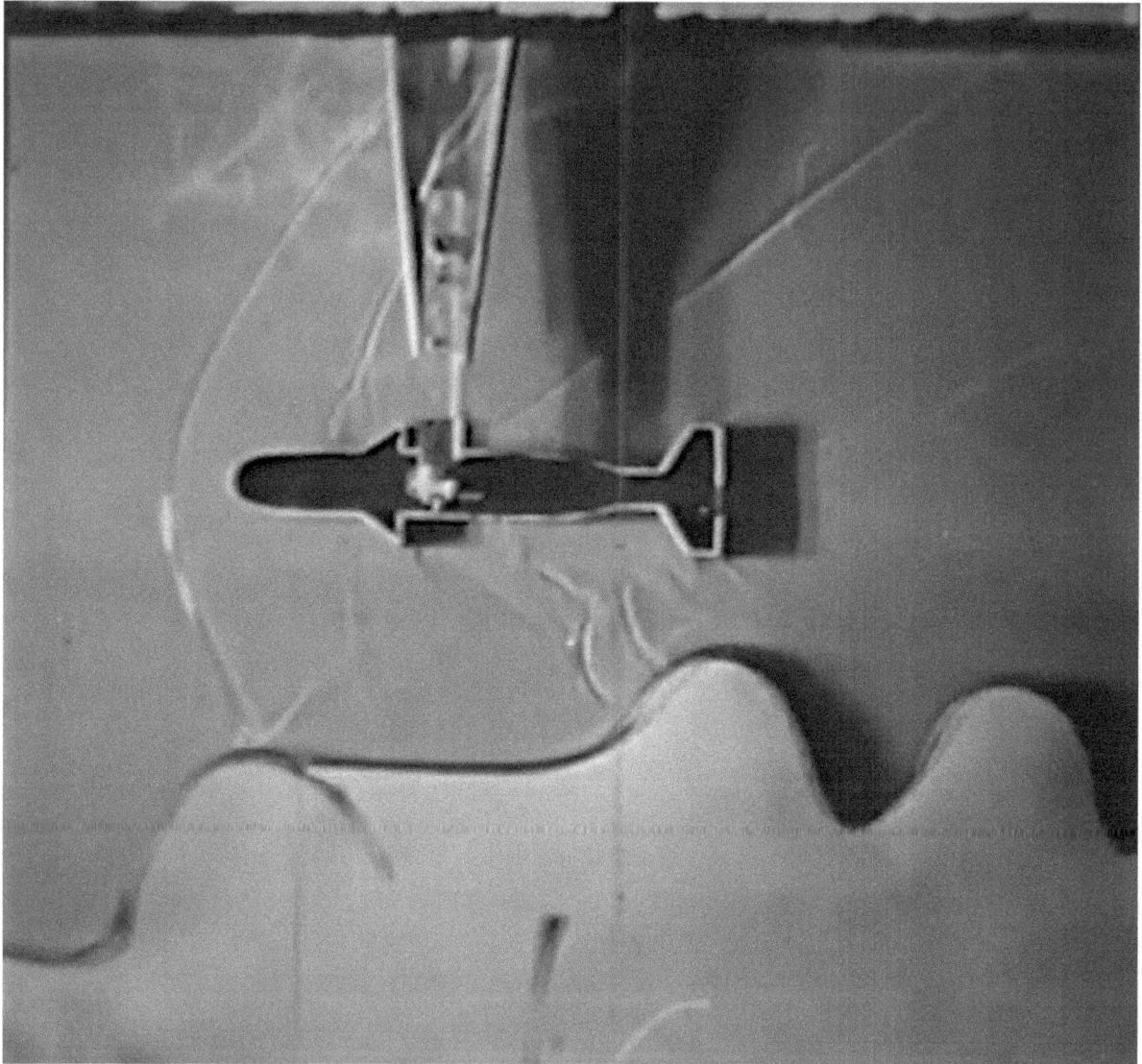

Photo 64

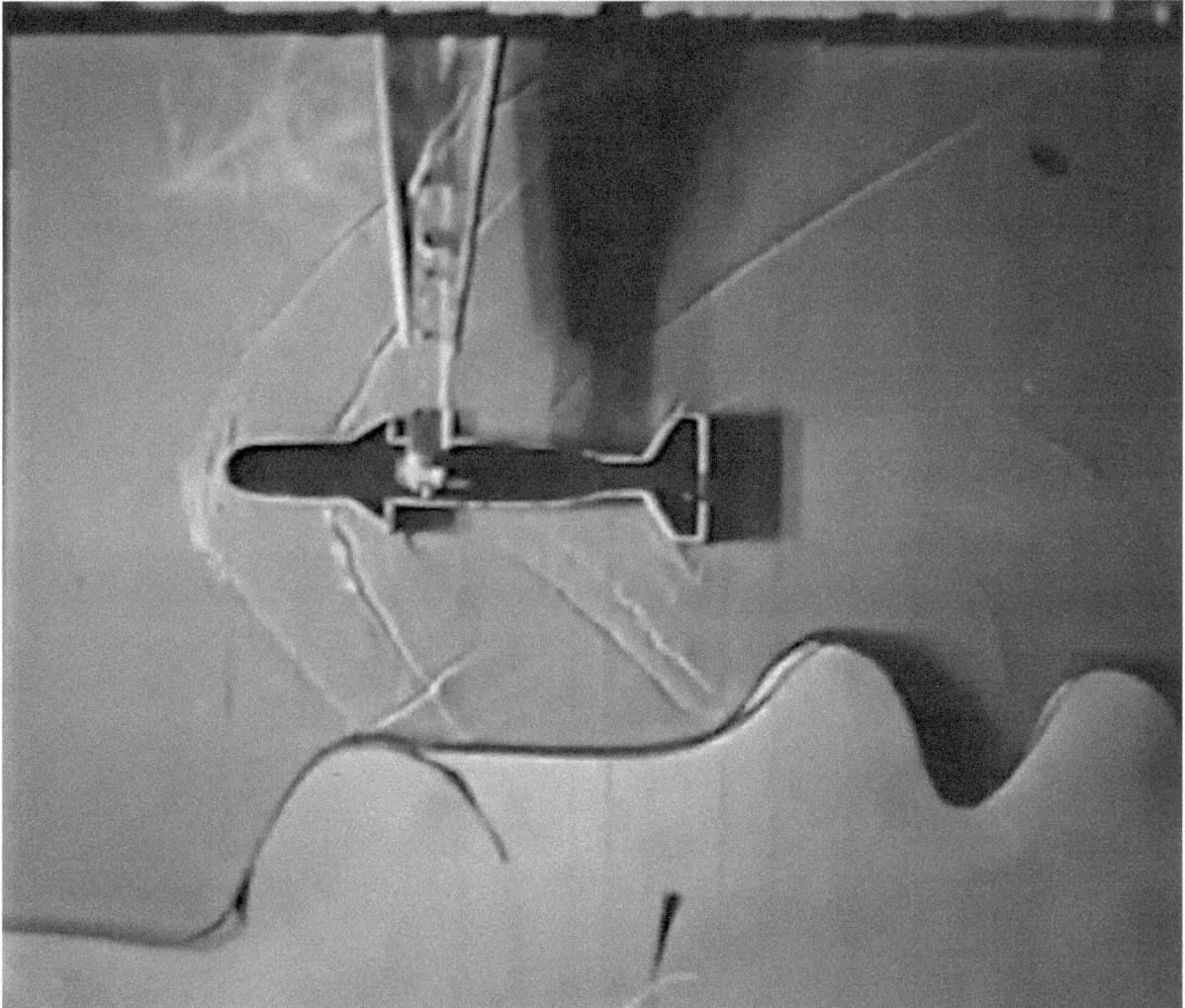

Photo 65

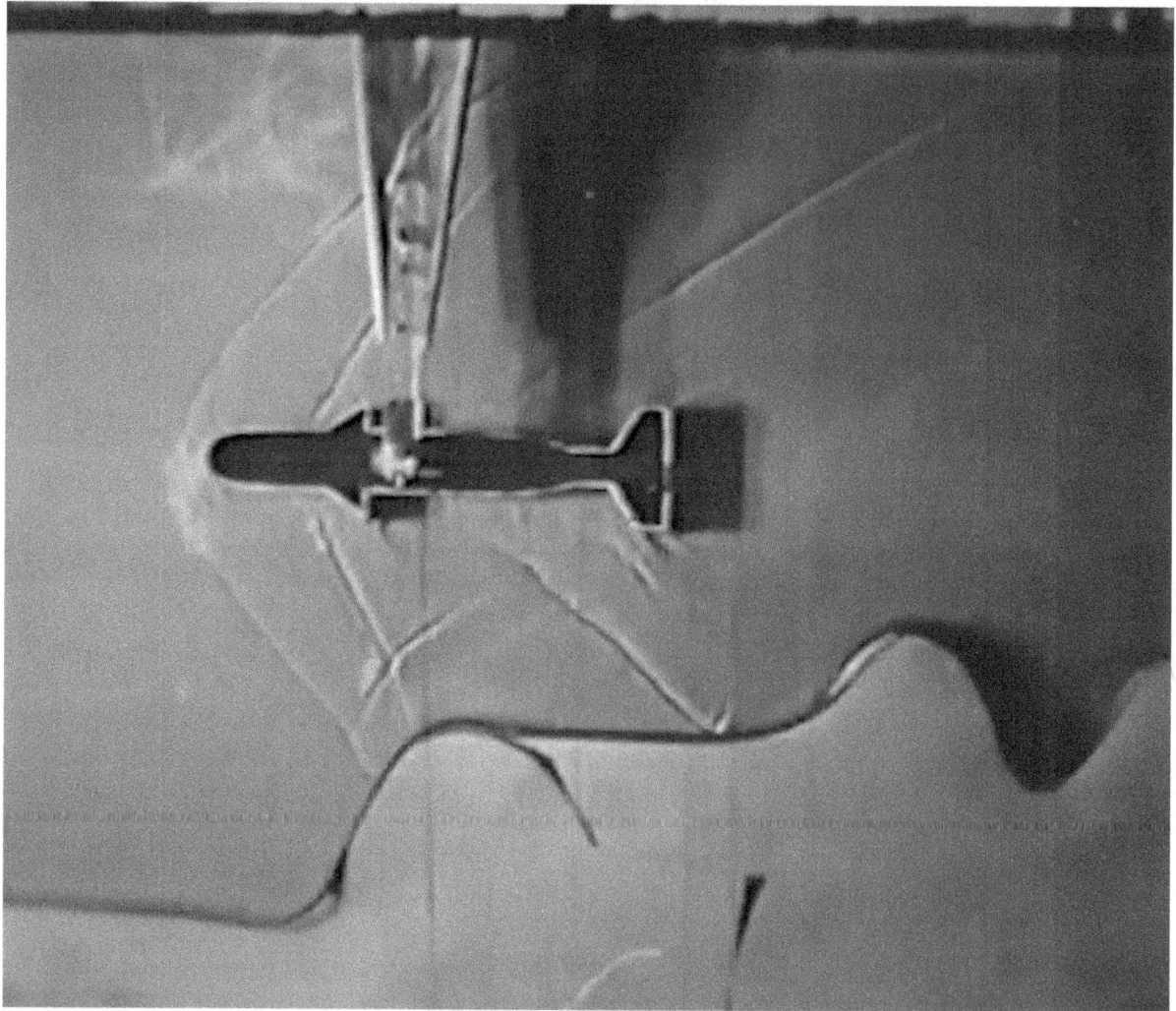

Photo 66

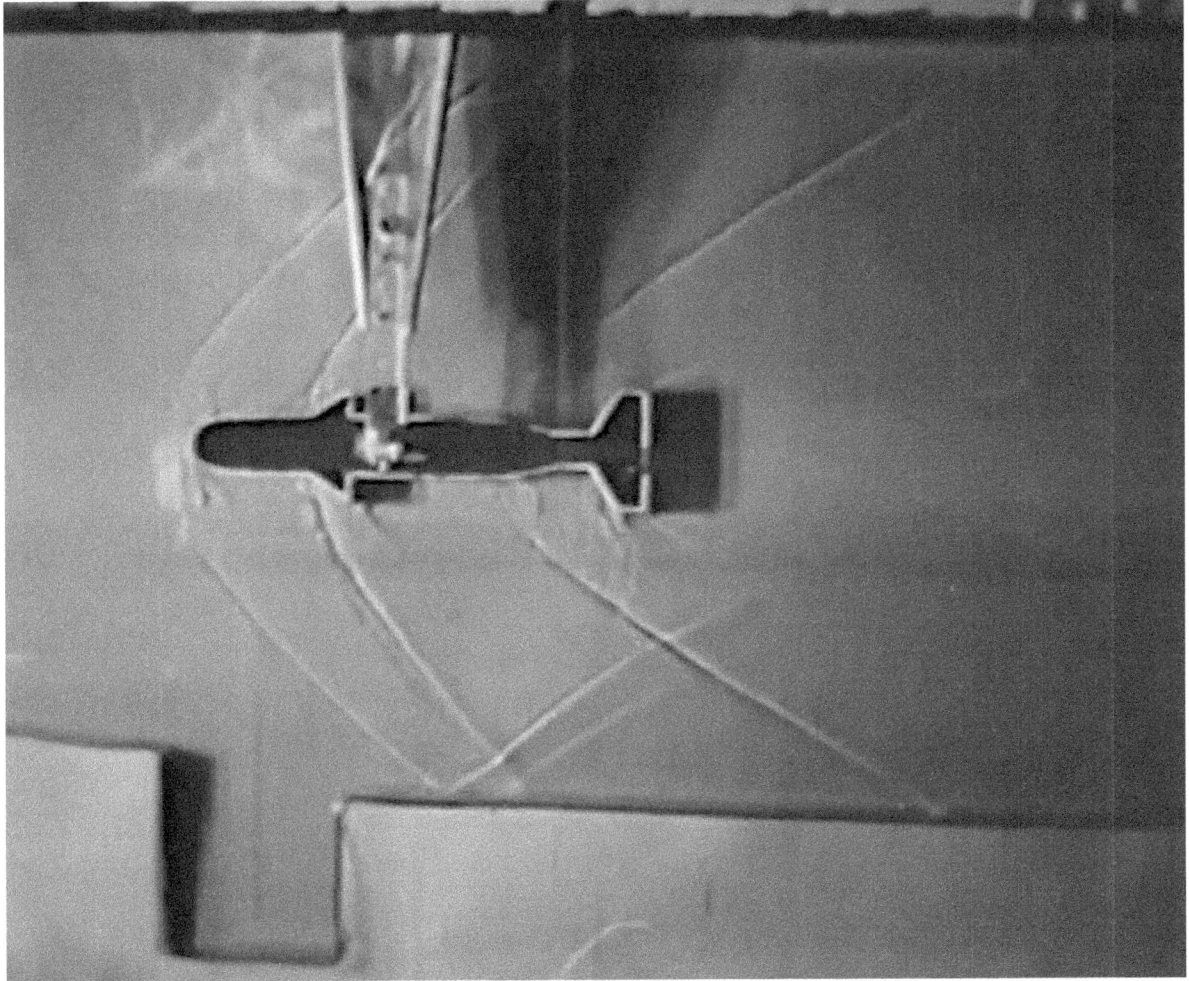

Photo 67

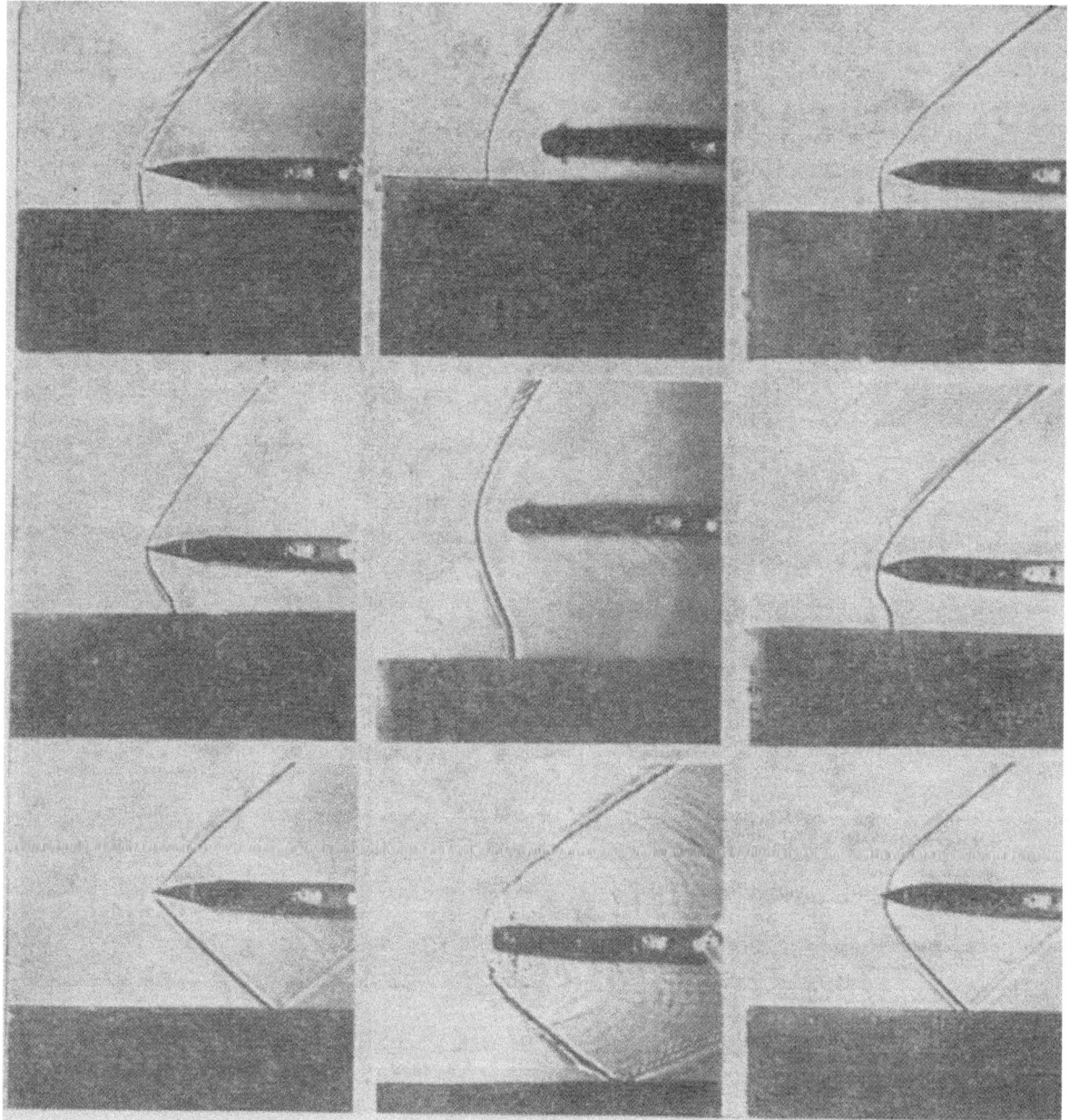

Photo 68

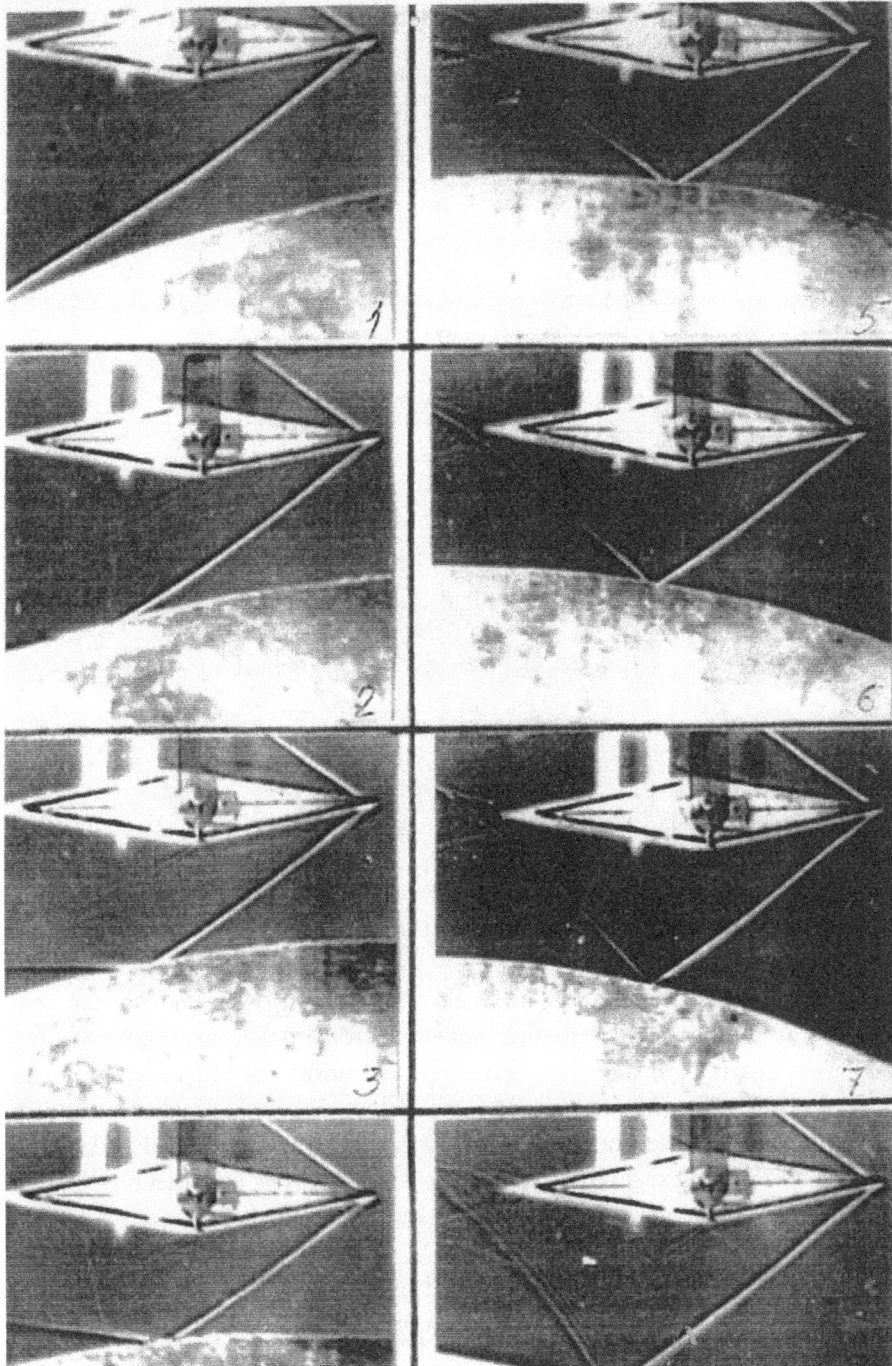

Photo 69
Film sequence showing a diamond-shaped airfoil traveling at supersonic speed in proximity to a convex surface.

Visualization of profile oscillations in supersonic flow

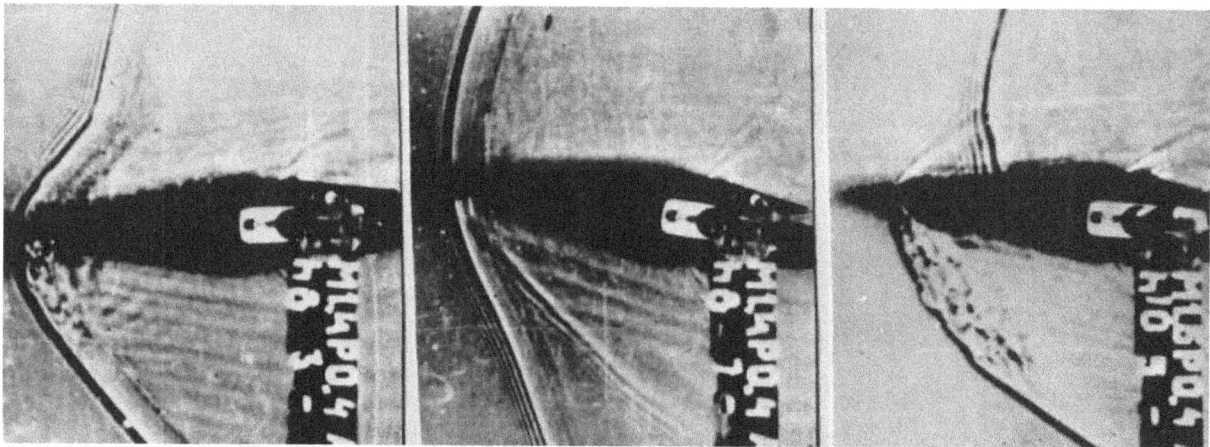

Photo 70
Flow anomalies in supersonic regime induced by oscillatory motion of an airfoil.

Due to the thousand-fold time deceleration in shallow water compared to gas, the gas-hydraulic analogy enables the modeling of practically relevant frequencies and amplitudes of body oscillations. This key advantage allowed the demonstration of pronounced shock wave deformations in supersonic flow ($M\infty = 1.3$–2.0) around an oscillating airfoil, in contrast to the stationary regime. The study revealed not only shock wave deformation but also the emergence of a new flow structure comprising discretely advancing shock waves and discretely lagging shock waves.

The left image shows shock wave deformation caused by an oscillating diamond-shaped airfoil in a supersonic flow. The center image captures a discretely advancing (catch-up) shock wave, while the right image depicts a discretely lagging shock wave.

Motion of Profiles along Complex Curvilinear Trajectories

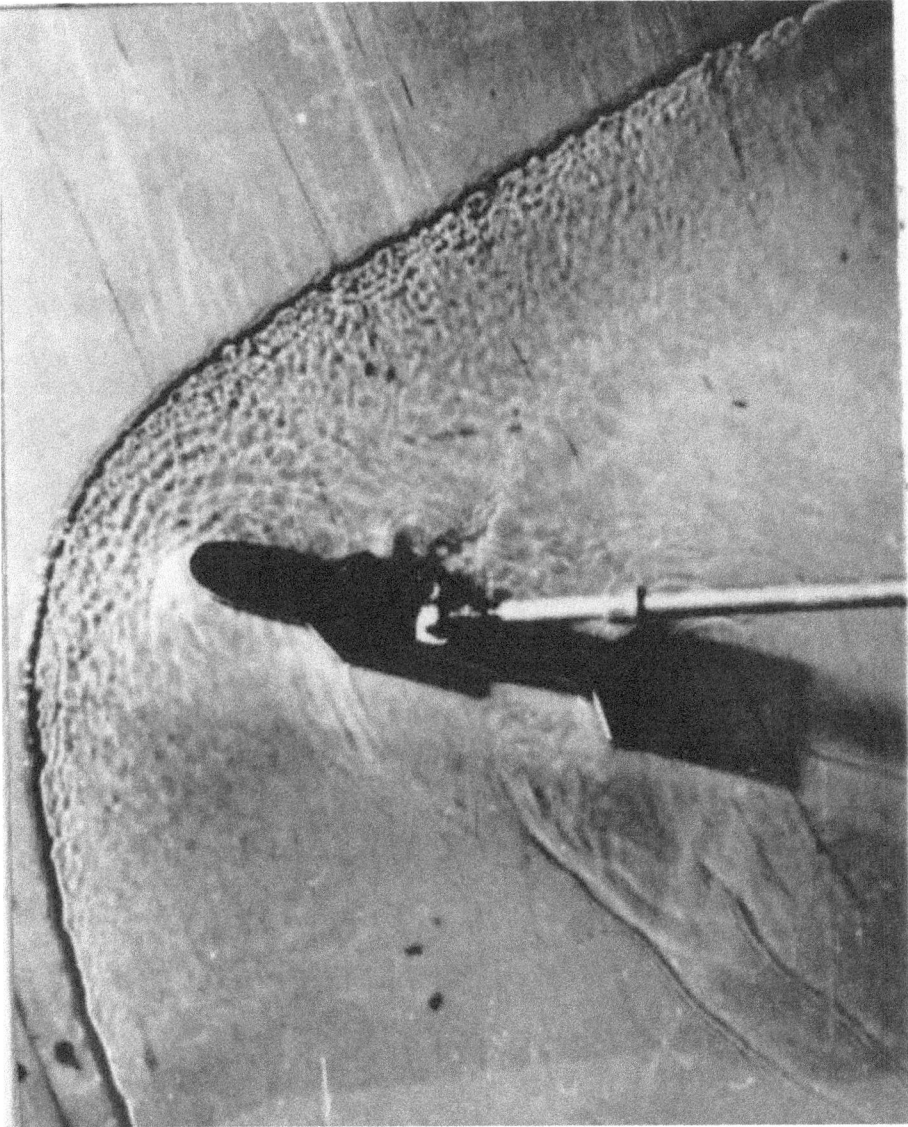

Photo 71

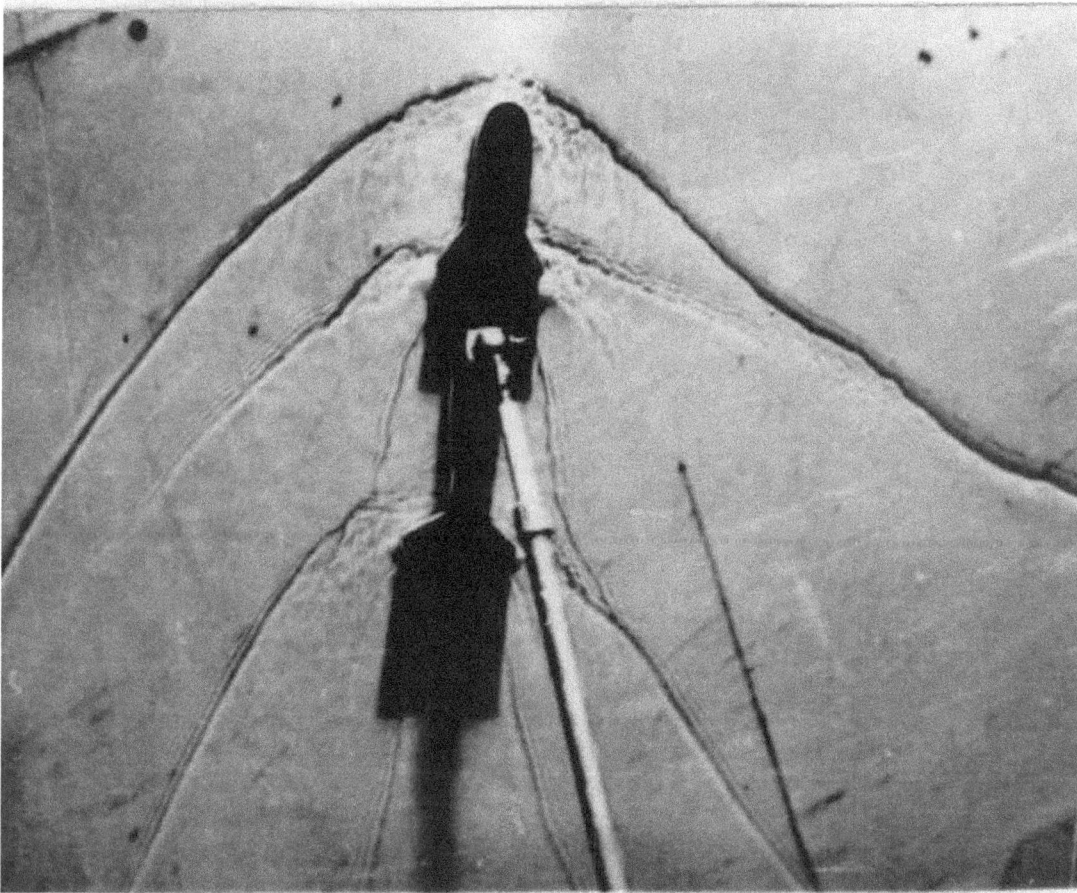

Photo 72

Photo 71, 72 Simulation of supersonic body motion along various curved trajectories (circular, elliptical, sinusoidal, etc.).

Flow Visualization Techniques

Application of Aluminum Powder for Flow Visualization in Gas-Hydraulic Analogy Experiments

Photo 73
Supersonic flow interaction with two tandem-aligned bodies undergoing separation, visualized using aluminum powder.

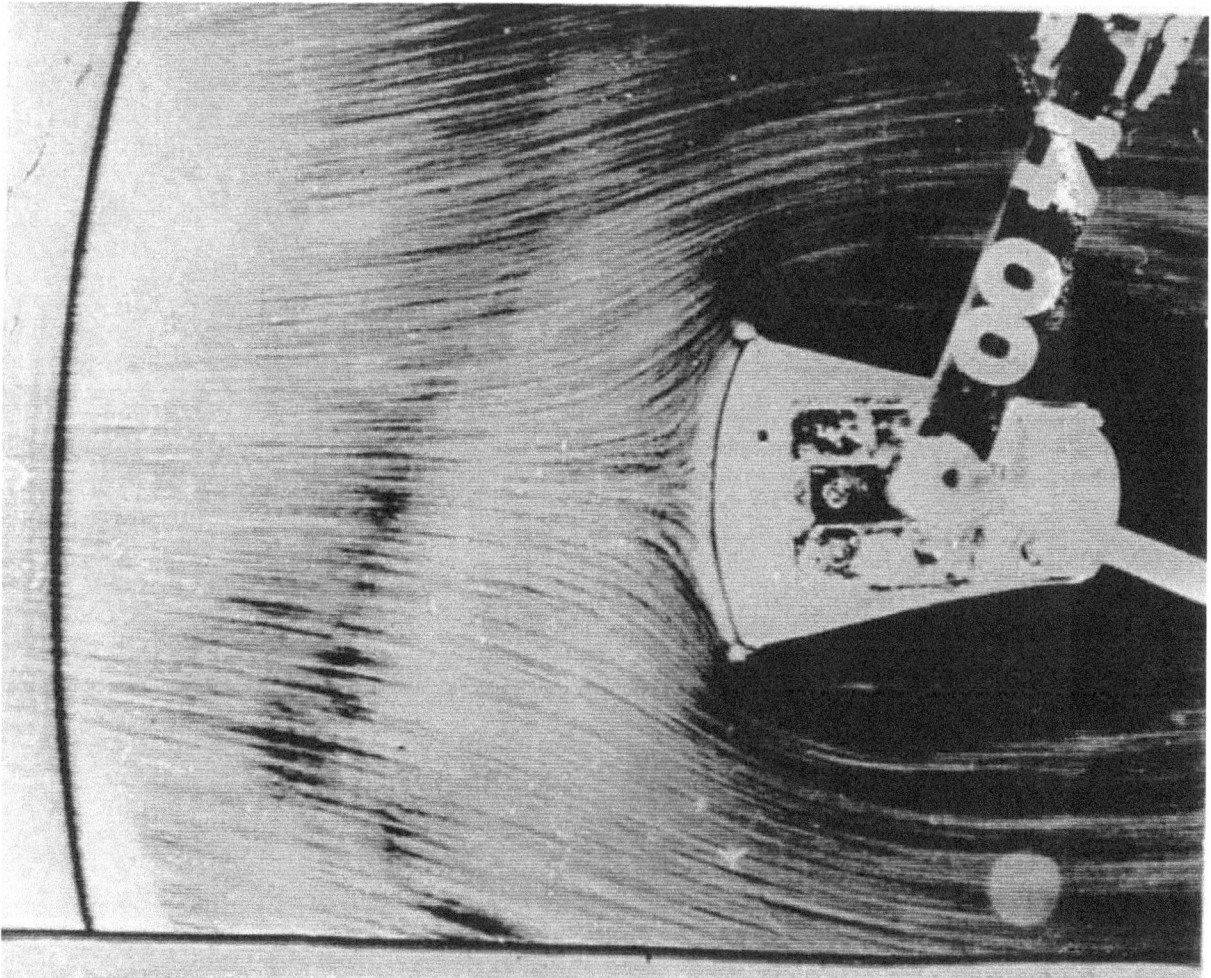

Photo 74
The use of aluminum powder enables shadow-based visualization of shock wave structures.

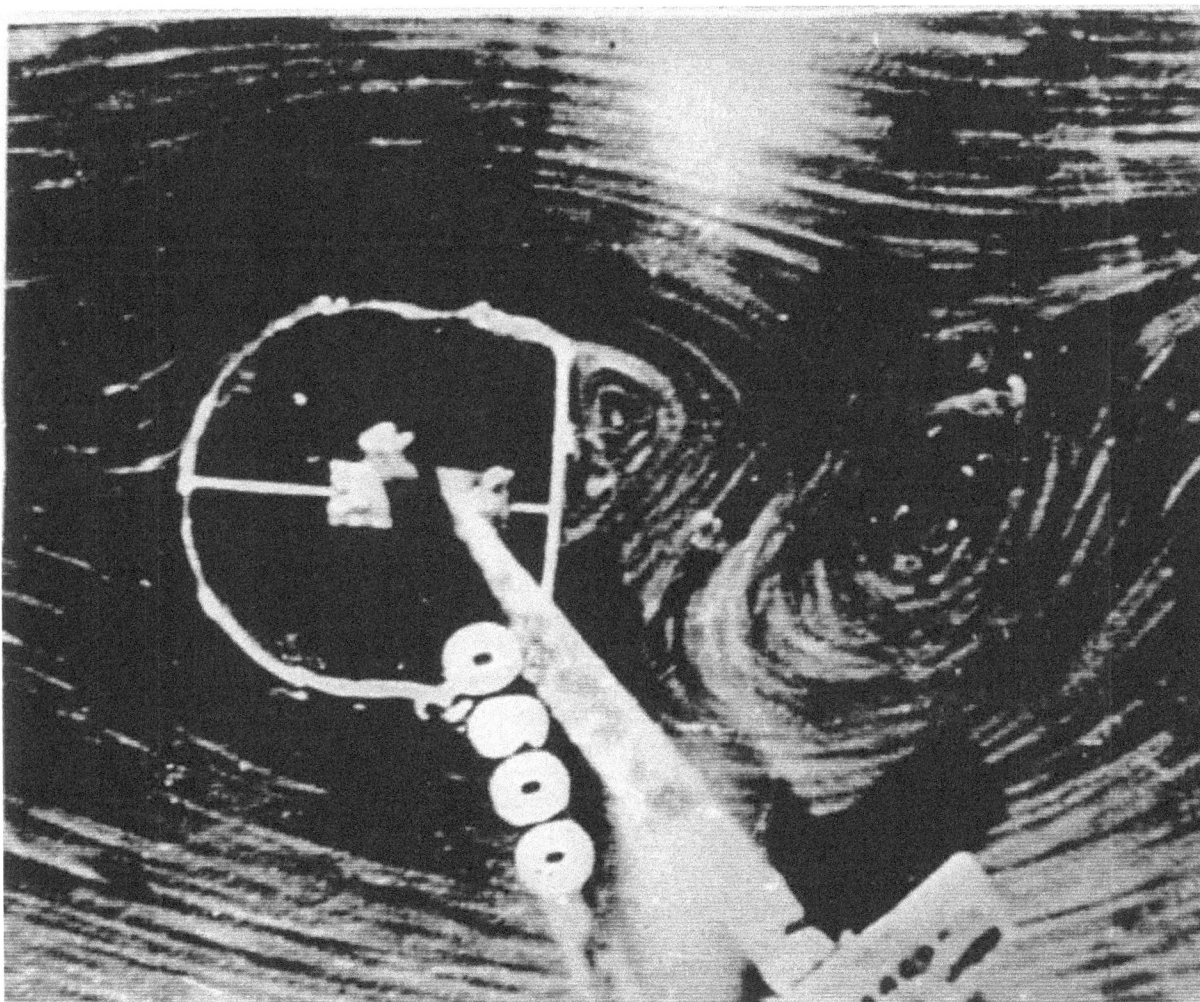

Photo 75
Visualization of Vortex Generation in the Inter-Body Flow Region

Bibliography

1. Arenze Z., Skew B.W. *Experimental and Numerical Study of Hydraulic Analogy to Supersonic Flow*. R & D Journal, 2008, 24(2), South African Institution of Mechanical Engineering.

2. Bomelburg H. *Die praktische Anwendbarkeit der Wasseranalogie in quantitativer Form auf spezielle Probleme der Gasdynamik*. Mitteilung aus der Max-Planck-Institut für Strömungsforschung, Nr.10, Göttingen, Selbstverlag, 1954, S.82.

3. Hafez M. *Hydraulic Analogy and Visualisation of Two-Dimensional Compressible Fluid Flows: Part 1: Theoretical Aspects*. Int. J. Aerodynamics, Vol. 6, No. 1, 2018.

4. Hafez M., Jan F., Linke B., Garretson I. *Hydraulic Analogy and Visualisation of Two-Dimensional Compressible Fluid Flows: Part 2: Water Table Experiments*. Int. J. Aerodynamics, Vol. 6, No. 1, 2018. https://www.researchgate.net/publication/323107944

5. Kazuyoshi Takayama. *Visualization of Shock Wave Phenomena*. Springer Nature Switzerland AG, 2019.

6. Laitone E.V. *A Study of Transonic Gas Dynamics by Hydraulic Analogy*. Journal of the Aeronautical Sciences, Vol. 19, No. 4, 1952, p. 265–272.

7. Merzkirch Wolfgang Flow Visualization. Institute for Thermo- and Fluid Dynamics Ruhr University, Bochum, West Germany. Academic Press, New York and London< 1974 A Subsidiary of Hartcourt Brace Jovanovich, Publishers,

8. Milton Van Dyke. *An Album of Fluid Motion*. The Parabolic Press, Stanford California, Fourth Printing, 1988.

9. Riaboushinsky D. *Sur l'analogue Hydraulique des mouvements d'un fluide compressible*. Comptes Rendus, Vol. 195, No. 22, pp. 998–999, Nov. 1932.

10. Samimy M., Breuer K.S., Leal L.G., Steen P.H. *A Gallery of Fluid Motion*. Cambridge University Press, 2003.

11. *Gallery of Fluid Motion*, Presented by the APS Division of Fluid Dynamics. https://gfm.aps.org/meetings/dfd-2023

12. Аэрофизические исследования сверхзвуковых течений: [Сборник статей / Отв. ред. д-р техн. наук Ю. А. Дунаев]; АН СССР. Физ.-техн. ин-т им. А. Ф. Иоффе, Москва–Ленинград, Наука, 1967, 302 стр.

13. Виноградов Р. И., Жуковский М. И., Якубов И.Р. *Газогидравлическая аналогия и её практическое применение*. Москва: Машиностроение, 1978.

14. Дитман А.О., Савчук В.Д., Якубов И.Р. *Методы аналогий в аэродинамике*. Москва: Машиностроение, 1987.

15. Гилинский М.М., Лебедев М. Г., Якубов И.Р. *Моделирование течений газа с ударными волнами*. Москва, 1984.

Aerodynamic man
Cartoon generated with ChatGPT during the album development process.

Joseph Yakubov

Born in 1938 in Tashkent, Uzbekistan (then USSR), Joseph Yakubov led pioneering research on the gas-hydraulic analogy at the Tashkent Polytechnic Institute named after Abu Rayhan Beruni from 1965 to 1995. He served as Professor, Doctor of Sciences, and Head of Department. He currently resides in Los Angeles, California, USA.

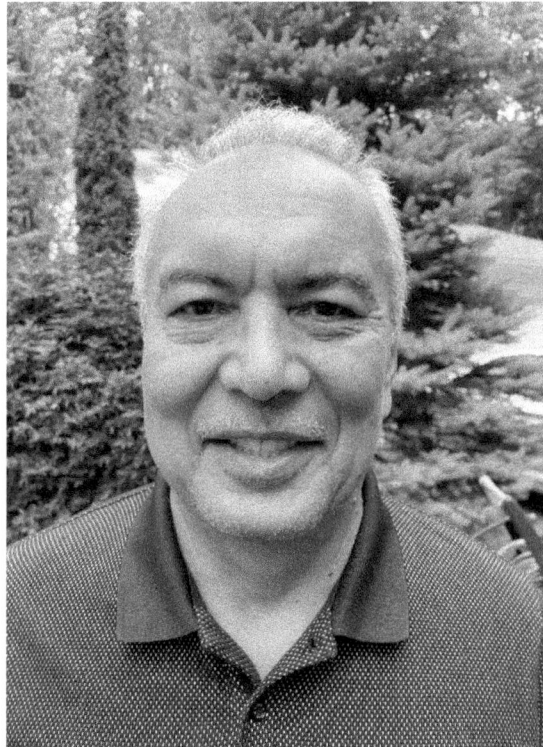

Bakhram Musabekov

Born in 1955 in Tashkent, Uzbekistan. A mechanical engineer in aircraft building and holding a PhD in fluid mechanics. From 1972 to 1980, he contributed to experimental research on the gas-hydraulic analogy method with Joseph Yakubov. His professional career spans multiple engineering institutions, including roles within the aerospace sector. Since 2022, he has been residing in Rhode Island, USA.

For inquiries regarding the gas-hydraulic analogy, please contact Dr. Bakhram Musabekov at: bakhram.musabekov@gmail.com